NASA SP-2000-4522

Flight Research: Problems Encountered and What They Should Teach Us

by
Milton O. Thompson

with
A Background Section by J.D. Hunley

NASA History Division
Office of Policy and Plans
NASA Headquarters
Washington, DC 20546

Monographs in
Aerospace History
Number 22
2000

Library of Congress Cataloging-in-Publication Data

Thompson, Milton O.
  Flight Research : problems encountered and what they should teach us / Milton O.
Thompson ; with a background section by J.D. Hunley.
    p. cm. – (Monographs in aerospace history ;) (NASA history series) (NASA SP-2000
  ; 4522)
 Includes bibliographical references and index.
  1. Aeronautics—Research—United States. 2. Airplanes—Flight testing. 3. High-speed
Aeronautics. I. Title. II. Series. III. Series: NASA history series IV. NASA SP ; 4522.

TL565.T46 2000
629.13'07'2073—dc2l   00-048072

Table of Contents

Foreword

The document by Milt Thompson that is reproduced here was an untitled rough draft found in Thompson's papers in the Dryden Historical Reference Collection. Internal evidence suggests that it was written about 1974. Readers need to keep this date in mind, since Milt writes in the present tense. Apparently, he never edited the document. Had he prepared it for publication, he would have done lots of editing and refined much of what he said.

I have not attempted to second guess what Milt might have done in revising the paper, but I have made some minor stylistic changes to make it more readable without changing the sense of what Milt initially wrote. Where I have qualified what Milt said or added information for the reader's benefit, I have done so either in footnotes or inside square brackets [like these]. The draft itself indicated that it should contain numerous figures to illustrate what he wrote, but no such figures were associated with the manuscript. I have searched out figures that appear to illustrate what Milt intended to show, but in some cases I have found none. When that has been the case, I have deleted his references to figures and simply kept his text, which does stand on its own.

For the most part, I have not attempted to bring his comments up to date, although in a few instances I have inserted footnotes that indicate some obvious changes since he wrote the paper. Despite—or perhaps because of—the paper's age, it offers some perspectives on flight research that engineers and managers not familiar with the examples Milt provides can still profit from in today's flight-research environment. For that reason, I have gone to the trouble to edit Milt's remarks and make them available to those who would care to learn from the past.

For readers who may not be familiar with the history of what is today the NASA Dryden Flight Research Center and of its predecessor organizations, I have added a background section. Those who do know the history of the Center may wish to skip reading it, but for others, it should provide context for the events Milt describes. Milt's biography appears at the end of the monograph for those who would like to know more about the author of the document.

Many people have helped me in editing the original manuscript and in selecting the figures. The process has gone on for so long that I am afraid to provide a list of their names for fear of leaving some important contributors out. A couple of them, in any event, requested anonymity. Let me just say a generic 'thank you' to everyone who has assisted in putting this document into its present form, with a special thanks to the Dryden Graphics staff members, especially Jim Seitz, for their work on the figures; to Jay Levine and Steve Lighthill for laying the monograph out; to Darlene Lister for her assistance with copy editing; and to Camilla McArthur for seeing the monograph through the printing process.

J. D. Hunley, Historian

Background: Flight Research at Dryden, 1946-1979

J. D. Hunley

Milt Thompson's account of lessons to be learned assumes some familiarity with the history of flight research at what was then called the Flight Research Center and its predecessor organizations—redesignated in 1976 the Hugh L. Dryden Flight Research Center. The following account provides a brief history of the

used in the war, had a top speed of 437 miles per hour when flying a level course at low altitude. This compared with low-level maximum speeds of 514 and 585 mph respectively for the Messerschmitt Me 262A and Gloster Meteor F.Mk. jet fighters, both of which thus still flew well below the

A P-51 Mustang on the lakebed next to the NACA High-Speed Flight Station in 1955. (NASA photo E55-2078)

subject that perhaps will provide a useful backdrop to what Milt had to say.

From the time of the Wright brothers' first flight in 1903 until the end of World War II, airplane technology evolved considerably. The early decades' mono- and biplanes of wooden framework, typically braced with wire and covered with cloth, gradually gave way to an all-metal construction and improved aerodynamic shapes, but most aircraft in World War II still featured propellers and even the fastest of them flew at maximum speeds of about 450 miles per hour. For example, the North American P-51 Mustang, one of the finest prop fighters

speed of sound (Mach 1) in level flight.[1]

Even so, during the early 1940s, airplanes like Lockheed's P-38 Lightning began to face the problem of compressibility in dives—characterized (among other things) by increased density, a sharp rise in drag, and disturbed airflow at speeds approaching Mach 1. The effects of compressibility included loss of elevator effectiveness and even the break-up of structural members such as the tail, killing pilots in the process. This problem was compounded by the absence of accurate wind-tunnel data

---

[1] See, e.g., Laurence K. Loftin, Jr., *Quest for Performance: The Evolution of Modern Aircraft* (Washington, DC, NASA SP-468, 1985), Chs. 1-5 and 9-10, esp. pp. ix, 7-45, 77-88, 128-136, 281-286, 484-490; Roger E. Bilstein, *Flight in America: From the Wrights to the Astronauts* (rev. ed.; Baltimore, MD.: Johns Hopkins Univ. Press, 1994), pp. 3-40, 129-145.

for portions of the transonic speed range in a narrow band on either side of the speed of sound.[2]

This situation led to the myth of a sound barrier that some people believed could not be breached. Since it appeared that jet aircraft would soon have the capability of flying in level flight into the transonic region—where the dreaded compressibility effects abound—a solution was needed for the lack of knowledge of transonic aerodynamics. A number of people (including Ezra Kotcher with the Army Air Forces [AAF] at Wright Field in Ohio, John Stack at the Langley Memorial Aeronautical Laboratory of the National Advisory Committee for Aeronautics [NACA] in Virginia, Robert Woods with Bell Aircraft, L. Eugene Root with Douglas Aircraft, and Abraham Hyatt at the Navy Bureau of Aeronautics) concluded that the solution could best result from a research airplane capable of flying at least transonically and even supersonically.

The emphases of these different organizations resulted in two initial aircraft—the XS-1 (XS standing for eXperimental Sonic, later shortened to X), for which Bell did the detailed design and construction for the AAF, and the D-558-1 Skystreak, designed and constructed by Douglas for the Navy. The XS-1 was the faster of the two, powered by an XLR-11 rocket engine built by Reaction Motors and launched from a B-29 or later a B-50 "mothership" to take full advantage of the limited duration provided by its rocket

A P-38 Lightning in flight in 1943. (NASA photo E95-43116-2)

[2] James O. Young, *Meeting the Challenge of Supersonic Flight* (Edwards AFB, CA: Air Force Flight Test Center History Office, 1997), pp. 1-2; John V. Becker, *The High-Speed Frontier: Case Histories of Four NACA Programs* (Washington, DC: NASA SP-445, 1980), esp. p. 95. It should be noted here that the first studies of compressibility involved tip speeds of propellers and date from 1918 to 1923. On these, see especially John D. Anderson, Jr., "Research in Supersonic Flight and the Breaking of the Sound Barrier" in *From Engineering Science to Big Science: The NACA and NASA Collier Trophy Research Project Winners*, ed. Pamela Mack (Washington, DC: NASA SP-4219, 1998), pp. 66-68. This article also provides excellent coverage of the early research of John Stack and his associates at the NACA's Langley Memorial Aeronautical Laboratory on the compressibility issue for aircraft (as opposed to propellers).

XS-1 Number 2 on the ramp at Edwards Air Force Base with its B-29 mothership. (NASA photo E-9)

propulsion. Stack and the other NACA engineers were skeptical about the rocket engine and less concerned about breaking the sound barrier than gathering flight data at transonic speeds, so they supported the AAF-Bell project with critical design data and recommendations but were more enthusiastic about the Navy-Douglas Skystreak. This was designed with an early axial-flow turbojet powerplant and was capable of flying only up to Mach 1. However, with comparably designed wings and a movable horizontal stabilizer recommended by the NACA, plus the ability to fly in the transonic region for a

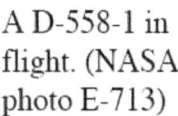

A D-558-1 in flight. (NASA photo E-713)

XS-1 Number 2 on Rogers Dry Lakebed in a photo that gives some sense of the expanse of natural runway provided by the dry lake. (NASA photo E49-001)

longer period of time, the D-558-1 complemented the XS-1 nicely and provided comparable data.[3]

To support the research flights, the contractors, the AAF (after September 1947, the Air Force), and the NACA sent teams of pilots and support personnel to the Muroc Army Air Field starting in September 1946 to support the XS-1, and then the Navy joined in to help fly the D-558. At Muroc, the 44-square-mile

XS-1 Number 1 in flight with copy of "Mach jump" paper tape data record of the first supersonic flight by Air Force Capt. Chuck Yeager. (NASA photo E-38438

---

[3] Young, *Supersonic Flight*, pp. 2-18; Becker, *High-Speed Frontier*, pp. 90-93; Richard Hallion, *Supersonic Flight: Breaking the Sound Barrier and Beyond, The Story of the Bell X-1 and the Douglas D-558* (rev. ed.; London and Washington, DC: Brassey's, 1997), esp. pp. 35-82; Louis Rotundo, *Into the Unknown: The X-1 Story* (Washington, DC and London: The Smithsonian Institution Press, 1994), esp. pp. 11-33; *Toward Mach 2: The Douglas D-558 Program*, ed. J. D. Hunley (Washington, DC: NASA SP-4222, 1999), esp. pp. 3-7.

NACA research aircraft on the ramp at the South Base area of Edwards Air Force Base, (left to right) D-558-2, D-558-1, X-5, X-1, XF-92A, X-4. (NASA photo EC-145)

Rogers Dry Lakebed provided an enormous natural landing field, and the clear skies and sparse population provided an ideal environment for conducting classified flight research and tracking the aircraft.[4]

The most immediate and dramatic result of these twin flight research efforts that proceeded simultaneously at Muroc was Air Force pilot Chuck Yeager's breaking the sound barrier on 14 October 1947 in the XS-1, for which feat he garnered the Collier Trophy the next year in conjunction with John Stack for the NACA and Larry Bell for his company.[5] The flight dispelled the myth about a sound barrier

In-flight photo of the X-3. (NASA photo E-17348)

---

[4] Rotundo, *Into the Unknown*, pp. 96, 123-132; James R. Hansen, *Engineer in Charge: A History of the Langley Aeronautical Laboratory, 1917-1958* (Washington, DC: NASA SP-4305, 1987), p. 297. Contrary to what is reported in a number of sources, the initial NACA contingent did not arrive on 30 Sept. 1946, with Walter C. Williams. Harold H. Youngblood and George P. Minalga arrived Sunday, 15 Sept., William P. Aiken, sometime in October after Williams' arrival. Telephonic intvws., Hunley with Youngblood and Aiken, 3 and 4 Feb. 1997.

[5] See especially Rotundo, *Into the Unknown*, pp. 279, 285.

X-2 in flight.
(NASA photo
E-2822)

and undoubtedly did much to gain credit for flight research, resulting in the small contingent of NACA engineers, pilots, and support people at Muroc becoming a permanent facility of the NACA and later NASA.

All of this was extremely important, but even more important than the record and the glory that went with it were the data that the NACA garnered from the flight research not only with the several X-1 and D-558-1 aircraft, but also with the Douglas D-558-2, the Bell X-2, the Douglas X-3 "flying stiletto," the Northrop X-4 semitailless, the Bell X-5 variable-sweep, and the Convair XF-92A delta-winged aircraft. Not all of these airplanes were successful in a conventional sense, even as research airplanes. But all of them provided important data for either validating or correcting information from wind tunnels and designing future airplanes ranging from the Century series of fighter aircraft to today's commercial transports, which still fly in the transonic speed range and feature the movable horizontal stabilizer demonstrated on the X-1 and D-558s. Even the ill-fated X-2, of which Dick Hallion has written, "its research was nil," and the X-3, which he has dubbed "NACA's glamorous hangar queen,"[6] nevertheless contributed to our understanding of the insidious problem of coupling dynamics. Furthermore, the Air Force-NACA X-2 program featured the first simulator used for the various functions of flight-test planning, pilot training, extraction of aerodynamic derivatives, and analysis of flight data.[7]

[6] Hallion, *On the Frontier: Flight Research at Dryden, 1946-1981* (Washington, DC: NASA SP-4303, 1984), pp. 78, 59 respectively. In both cases, Hallion's characterizations are justifiable in some degree.

[7] On the coupling and the computer simulation, Richard E. Day, *Coupling Dynamics in Aircraft: A Historical Perspective* (Dryden Flight Research Center, CA: NASA SP-532, 1997), esp. pp. 8-15, 34-36. On the value of the X-2 and X-3, see also *Ad Inexplorata: The Evolution of Flight Testing at Edwards Air Force Base* (Edwards AFB, CA: Air Force Flight Test Center History Office, 1996), pp. 14, 16. For the other research results, see especially Walter C. Williams and Hubert M. Drake, "The Research Airplane: Past, Present, and Future," *Aeronautical Engineering Review* (Jan. 1958): 36-41; Becker, *High-Speed Frontier*, pp. 42, 95-97; Hallion, *On the Frontier*, 59-62. In writing this account, I have benefited greatly from comments made to me over the years by long-time Dryden research engineer Ed Saltzman.

This NACA High-Speed Flight Station photograph of the Century Series fighters in formation flight was taken in 1957 (clockwise from left — F-104, F-101, F-102, F-100). (NASA photo E-2952)

Before this account discusses some of the other highlights of flight research at what became NASA's Dryden Flight Research Center, perhaps it should explain the differences and similarities between flight research and flight test. Both involve highly trained, highly skilled pilots and sometimes exotic or cutting-edge aircraft, although flight research can use quite old aircraft modified for particular kinds of research. There is no hard and fast dividing line separating flight research from flight test in practice, but flight research, unlike flight test in most applications, is oblivious to the particular aircraft employed so long as that airplane can provide the required flight conditions. On the other hand, flight test, as the name implies, often involves testing specific prototype or early production aircraft (somewhat later production aircraft in the case of operational flight-testing) to see if they fulfill the requirements of a particular contract and/or the needs of the user. In addition, however, flight testing—at least in the Air Force—involves flying aircraft that may be quite old to try to improve them and to develop their systems. For example, the Air Force Flight Test Center recently began testing the F-22, a brand new airplane, while at the same time it continued to test the F-15 and its systems even though various models of F-15s had been in the inventory for more than two decades.

In partial contrast to flight test, flight research sought and seeks fundamental understanding of all aspects of aeronautics, and in achieving that understanding, its practitioners may fly experimental aircraft like the early X-planes and the D-558s or armed service discards like early production models of the F-15s, F-16s, and F-18s researchers at Dryden are modifying and flying today. They may even fly comparatively new aircraft like the F-100 in its early days; here, however, the purpose is not to test them against contract standards but to understand problems they may be exhibiting in operational flight and learn of ways to correct them—a goal very similar to that

The D-558-2 Number 2 is launched from the P2B-1 in this 1956 NACA High-Speed Flight Station photograph. This is the same airplane that Scott Crossfield had flown to Mach 2.005 in 1953. (NASA photo E-2478)

of flight testing in its effort to improve existing aircraft. I should note in this connection that the Air Force Flight Test Center (as it is called today) and Dryden (under a variety of previous names) have often cooperated in flight research missions, with Air Force and NACA/NASA pilots flying together. So clearly flight test organizations in these cases participate in flight research just as research pilots sometimes engage in flight tests.[8] It should also be added that although many researchers at what is today Dryden might be quick, if asked, to point out the differences between flight test and flight research, many of them, including Milt in the account below, often used the two terms as if they were interchangeable.

To return to specific flight research projects at Dryden, on 20 November 1953, with NACA pilot Scott Crossfield in the pilot's seat, the D-558-2 exceeded Mach 2 in a slight dive, and on 27 September 1956, Air Force Capt. Mel Apt exceeded Mach 3 in the X-2 before losing control of the aircraft due to inertial coupling and plunging to his death.[9] With the then-contemporary interest in space flight, clearly there was a need at this point for research into hypersonic speeds (above Mach 5) and attendant problems of aerodynamic heating, flight above the atmosphere, and techniques for reentry. In early 1954, therefore, the NACA's Research Airplane Projects

---

[8] The account that comes closest to what I have said above is Lane Wallace's *Flights of Discovery: 50 Years at the NASA Dryden Flight Research Center* (Washington, DC: NASA SP-4309, 1996), pp. 4-8. On flight test per se, see *Ad Inexplorata,* esp. pp. 12-13. For a useful history of both flight testing and flight research, see Richard P. Hallion, "Flight Testing and Flight Research: From the Age of the Tower Jumper to the Age of the Astronaut," in *Flight Test Techniques,* AGARD Conference Proceedings No. 452 (copies of papers presented at the Flight Mechanics Panel Symposium, Edwards AFB, CA, 17-20 Oct. 1988), pp. 24-1 to 24-13. Finally, for an early discussion of flight research (despite its title) see Hubert M. Drake, "Aerodynamic Testing Using Special Aircraft," AIAA Aerodynamic Testing Conference, Washington, DC, Mar. 9-10, 1964, pp. 178-188. In writing and refining the above two paragraphs, I have greatly benefited from AFFTC Historian Jim Young's insightful comments about flight test, especially as it is practiced today at Edwards AFB, as well as from comments by Ed Saltzman. A point Ed offered that I did not incorporate in the narrative is that in the obliviousness of flight research to the specific aircraft used, it has more in common with wind-tunnel research than with flight test.

[9] Hallion, *On the Frontier,* pp. 308, 316.

The X-15 ship Number 3 (56-6672) is seen here on the lakebed at the Edwards Air Force Base, California. Ship Number 3 made 65 flights during the program, attaining a top speed of Mach 5.65 and a maximum altitude of 354,200 feet. (NASA photo E-7896)

Panel began discussion of a new research airplane that became the X-15. Developed under an Air Force contract with North American Aviation, Inc., and flown from 1959 to 1968, the X-15 set unofficial world speed and altitude records of 4,520 miles per hour (Mach 6.7) and 354,200 feet (67 miles).[10]

Much more importantly, however, the joint Air Force-Navy-NASA-North American program investigated all aspects of piloted hypersonic flight. Yielding over 765 research reports, the 199-flight program "returned benchmark hypersonic data for aircraft performance, stability and control, materials, shock interaction, hypersonic turbulent boundary layer, skin friction, reaction control jets, aerodynamic heating, and heat transfer,"[11] as well as energy management. These data contributed to the development of the Mercury,

In this photo the Number 1 XB-70A (62-0001) is viewed from above in cruise configuration with the wing tips drooped for improved controllability. (NASA photo EC68-2131)

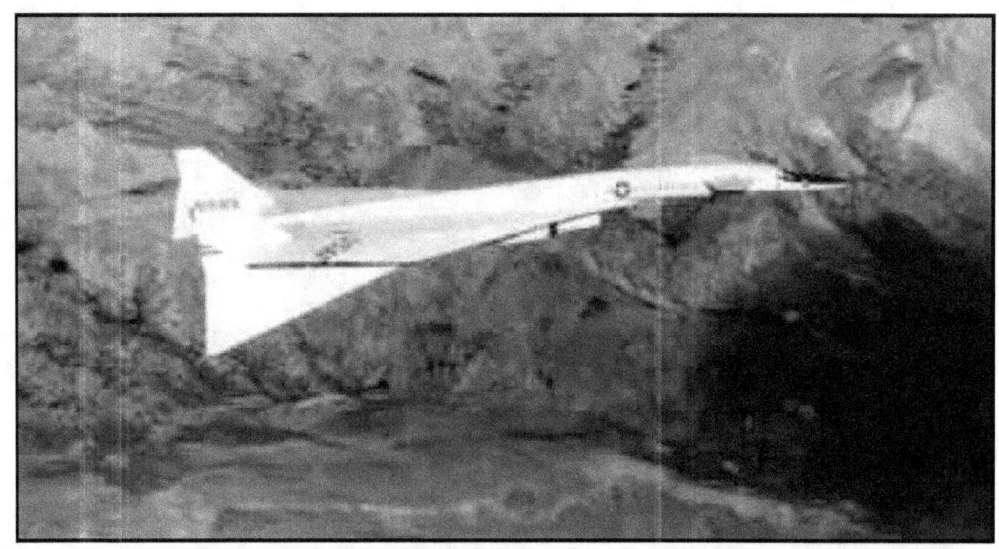

---

[10] See esp. *ibid.*, pp. 101-129, 333, 336, and Wendell H. Stillwell, *X-15 Research Results* (Washington, DC: NASA SP-60, 1965), p. vi and *passim*.

[11] Kenneth W. Iliff and Mary F. Shafer, *Space Shuttle Hypersonic Aerodynamic and Aerothermodynamic Flight Research and the Comparison to Ground Test Results* (Washington, DC: NASA Technical Memorandum 4499,1993), p. 2, for quotation and see also their "A Comparison of Hypersonic Flight and Prediction Results," AIAA-93-0311, paper delivered at the 31st Aerospace Sciences Meeting & Exhibit, Jan. 11-14, 1993, in Reno, NV.

Gemini, and Apollo piloted spaceflight programs as well as the later Space Shuttle program.[12]

Overlapping the X-15 program in time, the XB-70 also performed significant high-speed flight research. The XB-70 was the world's largest experimental aircraft. It was capable of flight at speeds of three times the speed of sound (roughly 2,000 miles per hour) at altitudes of 70,000 feet. It was used to collect in-flight information for use in the design of future supersonic aircraft, both military and civilian.

The more specific major objectives of the XB-70 flight research program were to study the airplane's stability and handling characteristics, to evaluate its response to atmospheric turbulence, and to determine the aerodynamic and propulsion performance. In addition,

there were secondary objectives to measure the noise and friction associated with airflow over the airplane and to determine the levels and extent of the engine noise during takeoff, landing, and ground operations. The first flight of the XB-70 was made on 21 September 1964. The Number two XB-70 was destroyed in a mid-air collision on 8 June 1966. Program management of the NASA-USAF research effort was assigned to NASA in March 1967. The final flight was flown on 4 February 1969. The program did provide a great deal of data that could be applied to a future supersonic transport or a large, supersonic military aircraft. It also yielded data on flight dynamics, sonic booms, and handling qualities.[13]

Another important high-speed flight research program involved the Lockheed YF-12 "Blackbird," precursor

A YF-12A in flight. (NASA photo EC72-3150)

---

[12] See, e.g., John V. Becker, "The X-15 Program in Retrospect," 3rd Eugen Sänger Memorial Lecture, Bonn, Germany, Dec. 4-5, 1968, copy in the NASA Dryden Historical Reference Collection; Milton O. Thompson, *At the Edge of Space: The X-15 Flight Program* (Washington, DC, and London: Smithsonian Institution Press, 1992); and the sources cited above.

[13] Hallion, *On the Frontier*, pp. 185-188; and see, e.g., P. L Lasagna and T. W. Putnam, "Engine Exhaust Noise during Ground Operation of the XB-70 Airplane" (Washington, DC: NASA TN D-7043, 1971) and C. H. Wolowicz and R. B. Yancey, *Comparisons of Predictions of the XB-70-1 Longitudinal Stability and Control Derivatives with Flight Results for Six Flight Conditions* (Washington, DC: NASA TM X-2881, 1973).

This photo shows the F-8 Digital-Fly-By Wire aircraft in flight. The project involving this aircraft contributed significantly to the flight control system on the space shuttles by testing and getting the bugs out of the IBM AP-101 used on the shuttles and by helping the Dryden Flight Research Center to develop a pilot-induced oscillation (PIO) suppression filter that reduced the likelihood of pilots overcontrolling the shuttles on landings and thereby creating excursions from the intended landing path. (NASA photo EC77-6988)

of the SR-71 reconnaissance airplane that flew at Dryden during the 1990s. Three YF-12s flew at Edwards in a joint NASA-AF research program between 1969 and 1979. The aircraft studied the thermal, structural, and aerodynamic effects of sustained, high-altitude, Mach 3 flight. They also studied propulsion, air flow and wind gusts, jet wake dispersion, engine stalls, boundary-layer noise, and much else. The 125 research reports the program produced contained vast amounts of information used in designing or improving other supersonic aircraft, including the SR-71. Among other things, engineers at Dryden developed a central airborne performance analyzer to monitor YF-12 flight parameters. It became the forerunner of the on-board diagnostic system used on the Space Shuttle.[14]

Not all of Dryden's flight research has concerned high-speed flight. One crucial flight research project that certainly had

implications for high-speed flight but was not restricted to that regime was the F-8 Digital Fly-By-Wire project. Dryden engineers replaced all purely mechanical linkages to flight-control surfaces (rudders, ailerons, elevators, and flaps) in an F-8C with electronic ones controlled by a digital flight-control system. Although there had been previous analog flight control systems, this was not only the first digital system but also the first electronic system without a conventional mechanical backup, using an analog backup instead. Flown in the 1970s and into the mid-1980s, the F-8 DFBW first used the Apollo computer developed by Draper Lab and then the IBM AP-101 later employed on the Shuttle. Flying this system without a mechanical backup was important in giving industry the confidence to develop its own digital systems since flown on the F-18, F-16, F-117, B-2, F-22, and commercial airliners like the

---

[14] On the YF-12, see esp. Berwin M. Kock, "Overview of the NASA YF-12 Program," *YF-12 Experiments Symposium*, Vol. 1, (3 vols.; Washington, DC: NASA CP-2054, 1978) plus more specialized papers in the volume; Robert D. Quinn and Frank V. Olinger, "Flight Temperatures and Thermal Simulation Requirements," *NASA YF-12 Flight Loads Program* (Washington, DC: NASA TM X-3061, 1974), pp. 145-183; and Hallion, *On the Frontier*, pp. 196-199, 349-356.

The Space Shuttle prototype Enterprise flies free after being released from NASA's 747 Shuttle Carrier Aircraft (SCA) during one of five free flights carried out at the Dryden Flight Research Center, Edwards, California, as part of the Shuttle program's Approach and Landing Tests (ALT). (NASA photo ECN-8611)

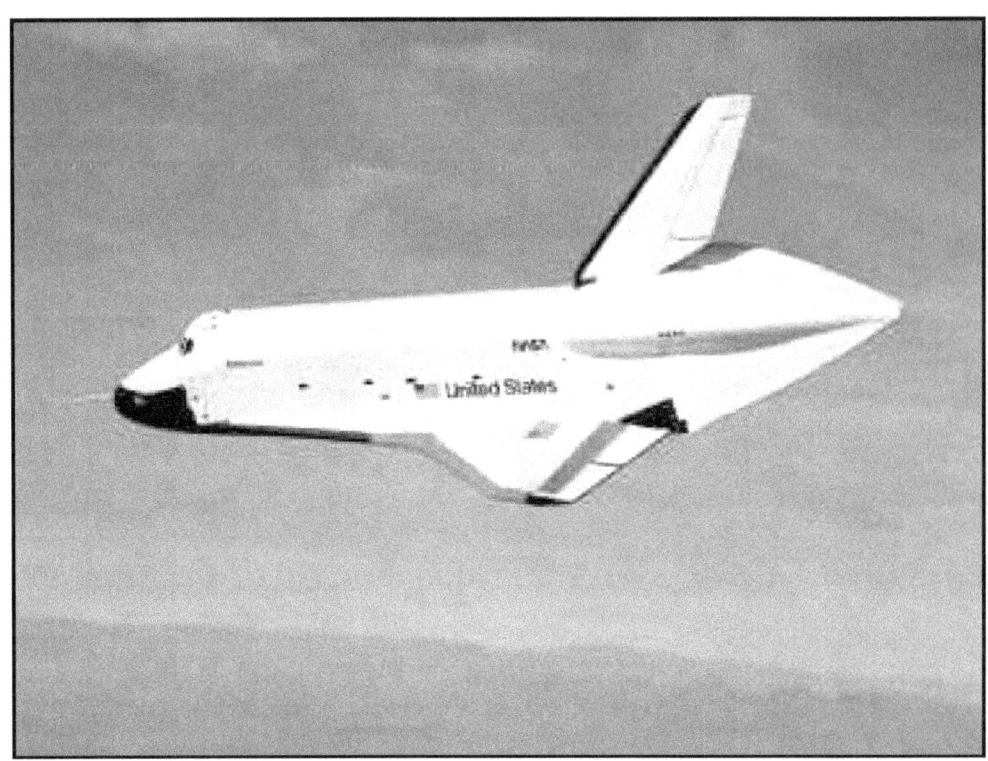

Boeing 777—not to mention the X-29 and X-31 research aircraft. Some of these aircraft would be uncontrollable without DFBW technology, which is not only lighter than mechanical systems but provides more precise and better maneuver control, greater combat survivability, and for commercial airliners, a smoother ride.[15]

While the F-8 Digital Fly-By-Wire project was still ongoing, Dryden hosted the Approach and Landing Tests for the Space Shuttle in 1977. These involved testing the 747 Shuttle Carrier Aircraft (SCA) modified to carry the Shuttle back to its launch location at the Kennedy Space Center in Florida (following Shuttle landings on the Rogers Dry Lakebed), plus flying the Shuttle prototype Enterprise mated to the 747, both without and with a crew on the Shuttle, and then five free flights of the Enterprise after it separated from the SCA, including four lakebed landings and one on the regular runway at Edwards. Flying without a tail-cone fairing around the dummy main engines as well as landing on the smaller runway for the first time, astronaut and former Dryden research pilot Fred Haise was keyed up and overcontrolled the orbiter on the concrete-runway landing, resulting in a pilot-induced oscillation. Once he relaxed his controls, the Enterprise landed safely, but only after some very uneasy moments.[16]

[15] James E. Tomayko, "Digital Fly-by-Wire: A Case of Bidirectional Technology Transfer," *Aerospace Historian* (March 1986), pp. 10-18, and *Computers Take Flight: A History of NASA's Pioneering Digital Fly-By-Wire Project* (Washington, DC: NASA SP-2000-4224, 2000); *Proceedings of the F-8 Digital Fly-By-Wire and Supercritical Wing First Flight's 20th Anniversary Celebration* (Edwards, CA: NASA Conference Publication 3256, 1996), Vol. I, esp. pp. 4, 15, 19-20, 34, 46-51, 56; oral history interview, Lane Wallace with Kenneth J. Szalai and Calvin R. Jarvis, Aug. 30, 1995, transcript in the Dryden Historical Reference Collection. See also Lane Wallace's account in *Flights of Discovery*, pp. 111-118.

[16] *Space Shuttle Orbiter Approach and Landing Test, Final Evaluation Report* (Houston, TX: NASA JSC-13864. 1978). For shorter, less technical descriptions, see Hallion, *On the Frontier*, pp. 242-250, and Wallace, *Flights of Discovery*, pp. 134-137.

This posed a hazard for Shuttle landings from space, because if a keyed-up pilot overcontrolled, the results might be more dangerous. There needed to be a correction to the Shuttle's flight control system. So the F-8 DFBW and other aircraft were pressed into service to find a solution. Dryden engineers suspected the problem lay in the roughly 270-millisecond time delay in the Shuttle's DFBW flight control system, so pilots flew the F-8 DFBW research airplane with increasing time delays to test this belief. When the experimental time-delay reached 100 milliseconds, Dryden research pilot John Manke was doing a touch-and-go landing-take-off sequence and entered a severe pilot-induced oscillation at a high angle of attack and low speed. As the aircraft porpoised up and down in increasingly severe oscillations, hearts stopped in the control room until fellow F-8 pilot Gary Krier reminded Manke to turn off the time delay, allowing him to climb to a safe altitude. The control room remained in a stunned silence until Gary keyed up the mike again and said, "Uh, John, I don't think we got any data on that; we'd like you to run it again." The ensuing laughter broke the tension. As a result of this and 13 other flights in March and April 1978, Dryden engineers had the data they needed to suppress control-surface action resulting from excessive pilot inputs. A suppression filter reduced the probability of a pilot-induced oscillation without affecting normal flying qualities, contributing to the safe landings of the Shuttle ever since.[17]

This 1964 NASA Flight Research Center photograph shows the Lunar Landing Research Vehicle (LLRV) Number 1 in flight at the South Base of Edwards Air Force Base. (NASA photo ECN-506)

---

[17] On this flight research, see especially Wallace's intvw. with Szalai and Jarvis; Tomayko, "Digital Fly-by-Wire," p. 17, and *Computers Take Flight*, pp. 113-114; and Wallace, *Flights of Discovery*, p. 137.

A much earlier contribution to the nation's space effort was the Lunar Landing Research Vehicle (LLRV). When Apollo planning was beginning in 1960, NASA began looking for a simulator to emulate the descent to the moon's surface. Three projects developed, but the most important was the LLRV developed by Bell Aerosystems in partnership with the Flight Research Center. Two LLRVs paved the way for three Lunar Landing Training Vehicles (LLTVs) supplemented by the LLRVs, which were converted into LLTVs.

Ungainly vehicles humorously called "flying bedsteads," they simulated the moon's reduced gravity on descent by having a jet engine provide five-sixths of the thrust needed for them to stay in the air. A variety of thrusters then handled the rate of descent and provided control. The vehicles gave the Apollo astronauts a quite realistic feel for what it was like to land on the Moon. Neil Armstrong said

that he never had a comfortable moment flying the LLTVs, and he crashed one of the LLRVs after it was converted to an LLTV, escaping by means of the ejection system. But he said he could not have landed on the Moon without the preparation provided by the LLTVs.[18]

Another very important contribution to the Shuttles and probably to future spacecraft came from the lifting bodies. Conceived first by Alfred J. Eggers and others at the Ames Aeronautical Laboratory (now the Ames Research Center), Mountain View, California, in the mid-1950s, a series of wingless lifting shapes came to be flown at what later became Dryden from 1963 to 1975 in a joint program with the Air Force, other NASA centers, and both Northrop and Martin on the industrial side. They included the M2-F1, M2-F2, M2-F3, HL-10, and X-24A and B. Flown at comparatively low cost, these low lift-over-drag vehicles

The HL-10 landing on the lakebed with an F-104 chase aircraft. (NASA photo ECN-2367)

[18] On the LLRVs and LLTVs, see Donald R. Bellman and Gene J. Matranga, *Design and Operational Characteristics of a Lunar-Landing Research Vehicle* (Washington, DC: NASA TN D3023, 1965) and Hallion, *On the Frontier*, pp. 140-146.

This photo shows the M2-F3 Lifting Body being launched from NASA's B-52 mothership at the NASA Flight Research Center. (NASA photo EC71-2774)

demonstrated both the viability and versatility of the wingless configurations and their ability to fly to high altitudes and then to land precisely with their rocket engines no longer burning. Their unpowered approaches and landings showed that the Space Shuttles need not decrease their payloads by carrying fuel and engines that would have been required for conventional, powered landings initially planned for the Shuttle. The lifting bodies also prepared the way for the later X-33 and X-38 technology demonstrator programs that feature lifting-body shapes to be used for, respectively, a potential next-generation reusable launch vehicle and a crew return vehicle from the International Space Station.[19]

A very different effort was the F-8 Supercritical Wing flight research project,

The X-24B landing on the lakebed with an F-104 safety chase aircraft. (NASA photo EC75-4914)

---

[19] For the details of this remarkable program, see R. Dale Reed with Darlene Lister, *Wingless Flight: The Lifting Body Story* (Washington, DC: NASA SP-4220, 1997); Milton O. Thompson with Curtis Peebles, *Flight without Wings: NASA Lifting Bodies and the Birth of the Space Shuttle* (Washington, DC: Smithsonian Institution Press, 1999).

The F-8 Supercritical Wing aircraft in flight. (NASA photo EC73-3468)

conducted at the Flight Research Center from 1971 to 1973. This project illustrates an important aspect of flight research at what is today Dryden because the design was the work of Dr. Richard Whitcomb at the Langley Research Center and resulted from his insights and wind-tunnel work. Frequently, projects flown at Dryden have resulted from initiatives elsewhere in NASA, in the armed services, in industry, or other places. However, researchers often discover things in flight that were only dimly perceived—or not perceived at all—in theoretical and wind-tunnel work, and flight research also can convince industry to adopt a new technology when it wouldn't do so as a result of wind-tunnel studies alone. In this case, Larry Loftin, director of aeronautics at Langley, said, "We're going to have a flight demonstration. This thing is so different from anything we've ever done before that nobody's going to touch it with a ten-foot pole without somebody going out and flying it."[20]

In this case, although there was some discovery resulting from the flight research—e.g., that there was some laminar flow on the wing that was not predicted, in addition to the numerous discrepancies Milt notes in his account below—generally there was good correlation between wind-tunnel and flight data. The SCW had increased the transonic efficiency of the F-8 by as much as 15 percent, equating to savings of $78 million per year in 1974 dollars for a 280-passenger transport fleet of 200-passenger airplanes. As a result of this study, many new transport aircraft today employ supercritical wings. Moreover, subsequent flight research with supercritical wings on the F-111 showed that the concept substantially improved a fighter aircraft's maneuverability and performance.[21]

A final project that should be mentioned here is the research with the three-eighths-scale F-15/Spin Research Vehicle. This was a sub-scale remotely piloted research vehicle chosen because of the risks

[20] Ted Ayers, "The F-8 Supercritical Wing; Harbinger of Today's Airfoil Shapes," *Proceedings of the F-8 . . . Supercritical Wing*, pp. 69-80, and Richard Whitcomb, "The State of Technology Before the F-8 Supercritical Wing," *ibid.*, pp. 81-92, quotation from p. 85.

[21] Ayers, "Supercritical Wing," p. 78; Whitcomb, "State of Technology," pp. 84, 90; Hallion, *On the Frontier*, pp. 202-208; Wallace, *Flights of Discovery*, pp. 90-92.

involved in spin testing a full-scale fighter aircraft. The remotely piloted research technique enabled the pilot to interact with the vehicle as he did in normal flight. It also allowed the flight envelope to be expanded more rapidly than conventional flight research methods permitted for piloted vehicles. Flight research over an angle-of-attack range of -20 degrees to +53 degrees with the 3/8-scale vehicle — during its first 27 flights through the end of 1975 in the basic F-15 configuration — allowed FRC engineers to test the math-

vehicle in other configurations at angles of attack as large as –70 degrees and +88 degrees.

There were 36 flights of the 3/8-scale F-15s by the end of 1978 and 53 flights by mid-July of 1981. These included some in which the vehicle — redesignated the Spin Research Vehicle after it was modified from the basic F-15 configuration — evaluated the effects of an elongated nose and a wind-tunnel-designed nose strake (among other modifications) on the airplane's stall/spin characteristics. Results of

This photograph shows NASA's 3/8th-scale remotely piloted research vehicle landing on Rogers Dry Lakebed at Edwards Air Force Base, California, in 1975. (NASA photo ECN-4891)

ematical model of the aircraft in an angle-of-attack range not previously examined in flight research. The basic airplane configuration proved to be resistant to departure from straight and level flight, hence to spins. The vehicle could be flown into a spin using techniques developed in the simulator, however. Data obtained during the first 27 flights gave researchers a better understanding of the spin characteristics of the full-scale fighter. Researchers later obtained spin data with the

flight research with these modifications indicated that the addition of the nose strake increased the vehicle's resistance to departure from the intended flight path, especially entrance into a spin. Large differential tail deflections, a tail chute, and a nose chute all proved effective as spin recovery techniques, although it was essential to release the nose chute once it had deflated in order to prevent an inadvertent reentry into a spin. Overall, remote piloting with the 3/8th-scale F-15 provided high-quality data about spin.[22]

[22] Kenneth W. Iliff, "Stall/Spin Results for the Remotely Piloted Spin Research Vehicle," AIAA Paper No. 80-1563 presented at the AIAA Atmospheric Flight Mechanics Conference, Aug. 11-13, 1980; Kenneth W. Iliff, Richard E. Maine, and Mary F. Shafer, "Subsonic Stability and Control Derivatives for an Unpowered, Remotely Piloted 3/8-Scale F-15 Airplane Model Obtained from Flight Test," (Washington, DC: NASA TN D-8136, 1976).

In these and many other projects, what is today the Dryden Flight Research Center has shown that while theory, ground research facilities, and now Computational Fluid Dynamics are critical for the design of aircraft and for advancing aeronautics, flight research is also indispensable. It serves not only to demonstrate and validate what ground research facilities have discovered but also—in the words of Hugh Dryden—to "separate the real from the imagined . . ." and to discover in flight what actually happens as far as instruments and their interpretation will permit.[23] This essential point is reemphasized in Milt's study from his own particular perspective, but his account also contains a great deal more that practitioners of flight research today—and perhaps even ground researchers—would do well to heed.

Portrait of Dr. Hugh L. Dryden a couple of years after he made the remark quoted in the narrative. (NASA photo E-4248)

---

[23] For the quotation, Hugh L. Dryden, "General Background of the X-15 Research-Airplane Project," in the NACA, *Research-Airplane-Committee Report on Conference on the Progress of the X-15 Project* (Langley Field, VA: Compilation of Papers Presented, Oct. 25-26, 1956): xix. Dryden's comment related specifically to the X-15 but has more general applicability. On the need for interpretation of data from instruments, see Frederick Suppe's interesting "The Changing Nature of Flight and Ground Test Instrumentation and Data: 1940-1969" on the Internet at http://carnap.umd.edu:90/phil250/FltTest/FltTest1.pdf. Of course, with the use of lasers in a variety of applications today to augment more traditional instrumentation, and with careful calibration of instruments as well as the use of instruments from different manufacturers in the same general location on an aircraft, there is less room for assumption and interpretation as well as for theoretical models to bias the understanding of flight research data than otherwise would be the case. But whenever aeronautical researchers use instruments in an experimental environment, there is always a need to spend a lot of time understanding what those instruments measure and how they do it to ensure accuracy in using data from them.

Problems Encountered in Flight Research

Milt Thompson

Introduction

The NASA Flight Research Center (FRC—formerly the NACA High-Speed Flight Station [and now known as the NASA Dryden Flight Research Center]) has been involved in experimental and research flight testing for over 27 years. FRC's experience began with the X-1 series of aircraft and extended through all the manned X-series aircraft, the D-558 series, and most recently the lifting bodies. Other experience was also gained with unusual vehicles such as paragliders and the Lunar Landing Research Vehicles. FRC has flight tested vehicles with operating speeds ranging from zero to 4,500 miles per hour and altitude ranges from ground level to 354,000 feet. Over 5,000 research flights have been made in over 60 different types of research aircraft. Only three aircraft and two pilots have been lost during research testing and none of these losses were attributable to negligence or inadequate planning or preparation.[24]

This is a remarkable record, especially considering the extremely hazardous nature of the testing FRC has been involved in. FRC has, however, had a number of accidents and incidents not involving the loss of an aircraft or a pilot. Numerous problems have been encountered in flight that were unpredicted or unanticipated. This, of course, is the justification for flight-testing. This document will describe some typical examples of the kinds of problems we have encountered. The intent is to make people aware of the kinds of problems we have encountered so that these same mistakes will not be repeated as they have been so often in the past.

The kinds of problems that we have encountered can be categorized into hardware problems, aerodynamic problems, and what might be called environmental problems. Hardware problems are those where a component or subsystem does not perform up to expectations. The component or subsystem doesn't function properly or fails completely. Aerodynamic problems are those encountered because the wind-tunnel predictions were not accurate or were misinterpreted or even inadequate. Environmental problems are those that show up only in flight. They generally result from a lack of foresight or understanding of the effects of the environment on a subsystem or component, or the vehicle itself.

Of the three types of problems, the emphasis will be on aerodynamic- and environmental-type problems. Two research aircraft have been selected as the prime examples, the HL-10 and the X-15. The HL-10 was an unconventional configuration with state-of-the-art off-the-shelf subsystems. Its problems, as you might suspect, were aerodynamic in nature. The X-15 was a relatively conventional configuration but most of its subsystems were newly developed and many pushed the state of the art. Its problems were mainly with subsystems. Both vehicles explored new flight regimes.

Aerodynamic Problems

Figure 1 illustrates an example of a serious problem encountered on the first flight of the HL-10—flow separation. The flow separation occurred at the junction of the tip fin and the fuselage. It occurred in flight as the pilot began his practice flare at altitude. When this occurred, the pilot essentially lost all pitch and roll control.

---

[24] Milt did not specify, but presumably he meant Howard Lilly's crash after takeoff due to compressor disintegration in the D-558-1 No. 2 on 3 May 1948, which resulted in Lilly's death; the crash of the M2-F2 without loss of life on 10 May 1967; and Michael Adams' fatal accident in X-15 No. 3 on 15 Nov. 1967. Although badly damaged, the M2-F2 was not lost and was rebuilt with a center fin to make it more stable and a more successful research airplane. This list does not include the deaths of Air Force Maj. Carl Cross and NASA pilot Joe Walker as a result of a mid-air collision between an XB-70A and an F-104N in 1966 because that did not occur as part of a research flight.

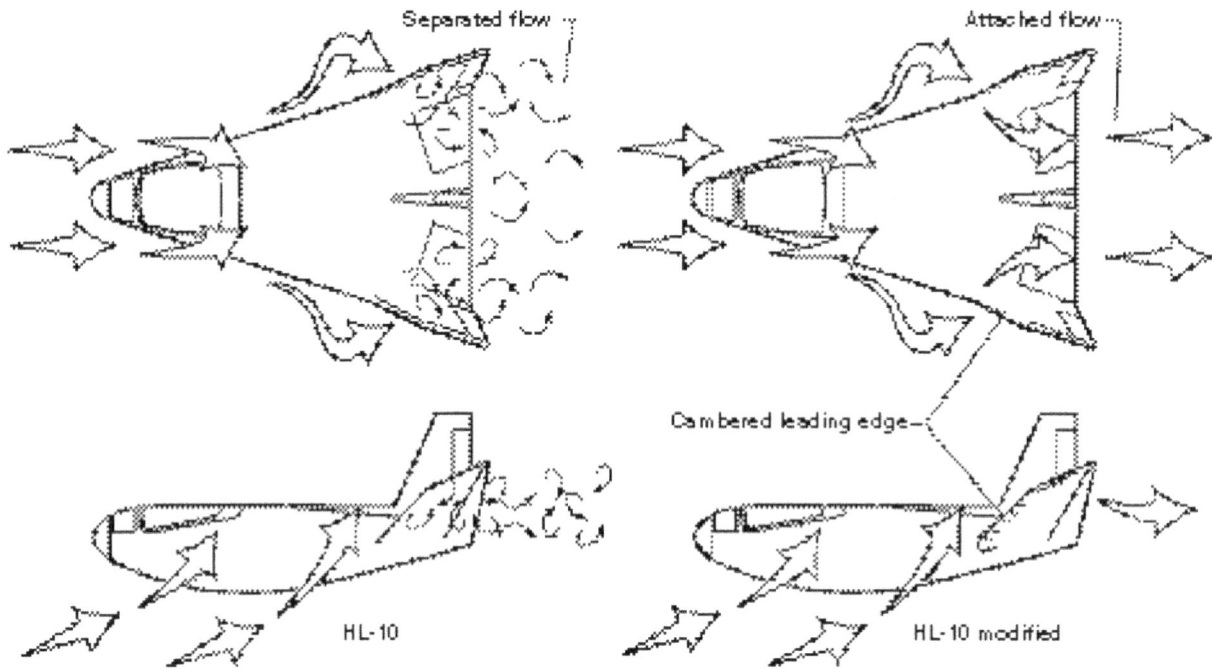

He had almost full right aileron in, and he was rolling slowly to the left. In this case, we were lucky. The vehicle recovered by itself since the flow separation also caused a nose-down pitching moment that lowered the angle of attack, causing the flow to reattach. A detailed reassessment of the wind-tunnel data revealed some slight evidence of a potential separation problem at the flight conditions that produced it; however, a substantial amount of additional wind-tunnel testing was required to confirm this and define a fix.

On that same flight we had longitudinal-control-system limit-cycle and sensitivity problems. The pilot used only one inch of longitudinal stick deflection from flare initiation at 300 knots to touchdown at 200 knots. The sensitivity and control-

system limit-cycle problems were primarily a result of the elevon effectiveness being higher than anticipated. I say "anticipated" rather than "predicted" because the measured effectiveness compared quite well with that measured in the small-scale wind tunnel; however, we had chosen to believe the full-scale wind-tunnel results. This is an interesting case since the full-scale wind-tunnel data were obtained using the actual flight vehicle as the model, and the Reynolds number range was from 20 to 40 million. The small-scale model was a 0.063-scale model (16 inches long), and the Reynolds number range was an order of magnitude lower—2 to 4 million. Flight Reynolds numbers ranged from 40 to 80 million.[25]

One might question whether it could be a compressibility effect, but one wouldn't

Figure 1: Schematic showing flow separation in the original HL-10 design and the modification to the HL-10 that solved the problem. This is not the exact illustration Milt intended for this paper, but it shows what he was talking about. (Original drawing by Dale Reed; digital version by the Dryden Graphics Office).

---

[25] Reynolds number, named after Osborne Reynolds, is a non-dimensional parameter equal to the product of the velocity of, in this case, an airplane passing through a fluid (air in this instance), the density of the fluid, and a representative length, divided by the fluid's viscosity. In shorthand, this is the inertial forces divided by the viscous forces for the mass of air acted upon by the vehicle. Among other uses, it served to compare data from wind-tunnel models with that from full-sized airplanes or components. The Reynolds number was not determined solely by the viscosity of the air. A large transport aircraft, for example, would have a much larger Reynolds number when flying through air at a given altitude, location, and time than would a small model simply because of the difference in size and the amount of air displaced. Furthermore, the Reynolds number would be much larger at the rear of a flight vehicle than at the front.

expect compressibility effects at 0.4 Mach number. We did, however, see compressibility effects as low as Mach 0.5 on the X-24A. It should be noted that flight-measured longitudinal stability was higher than predicted by either the small-scale or full-scale tunnels, whose data agreed quite well.

In the case of aileron characteristics, again the small-scale and full-scale tunnel results agreed quite well; however, the flight-measured results were higher than either, and again we had control-system limit-cycle and sensitivity problems during flight. The predicted subsonic longitudinal trim was off by approximately four degrees in angle of attack due to a combination of discrepancies in zero-lift pitching moment as well as static stability and control effectiveness. Discrepancies in longitudinal trim of roughly this same magnitude were observed in each of the lifting bodies.

Figure 2: Controllability boundaries for the X-24B at Mach 0.95. [A]

The HL-10 configuration had over 8,000 hours of wind-tunnel testing. One model that was tested was actually larger than the flight vehicle—28 feet long as com-

pared to a 20-foot flight vehicle, or 1.4 scale. The actual flight vehicle was tested in the 40X80-foot tunnel at Ames Research Center. You couldn't get better model fidelity, and yet we still saw discrepancies between the predicted and flight-measured data.

Aerodynamic discrepancies were not restricted to the HL-10 configuration. The HL-10 was simply used as an example. Each of the other lifting bodies exhibited similar kinds of discrepancies between predicted and flight data. The M2-F2 wind-tunnel tests were conducted and analyzed by another team of experts including people such as [Alfred J.] Eggers, [Clarence] Syvertson, [Jack] Bronson, [Paul F.] Yaggy, and many others, and yet again, the predictions were not perfect. The X-24A configuration was developed and tested by the Martin Company for the United States Air Force (USAF). It was a highly optimized and finely tuned configuration. The X-24A designers, for example, detected the potential for a flow separation problem at the fin-fuselage juncture and tested over twenty different fin leading-edge configurations before settling on the final leading edge for the flight vehicle. As meticulous as these designers were, we still saw some slight evidence of unpredicted flow separation.

On the X-24A, we also observed a discrepancy in aileron yawing-moment derivative. In terms of the actual numerical value, the discrepancy was small. In terms of percentage, it was an error by a minus 100 percent. In terms of vehicle handling qualities, the discrepancy was

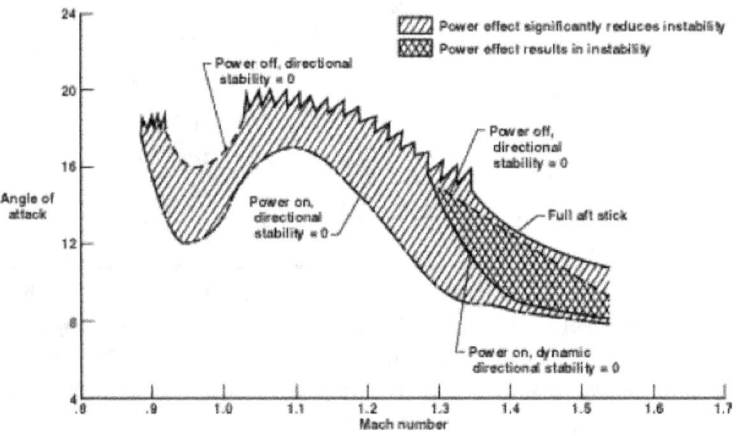

[A] Adapted and simplified from Christopher J. Nagy and Paul W. Kirsten, "Handling Qualities and Stability Derivatives of the X-24B Research Aircraft" (Edwards AFB, CA: AFFTC-TR-76-8, 1976), p. 56. It is obvious that this was not precisely the figure Milt had in mind, but it illustrates his point. Note that Nagy and Kirsten comment on p. 54, "Although modeling of the rocket exhaust conditions was not exact (hence the resulting data was not considered to be accurate), the results were used as guidelines to evaluate the potential loss of stability with the rocket engine on." They added, "The comparison of the handling qualities boundaries before and after the flight-test program exemplifies the need for an incremental envelope expansion approach to flight test of new aircraft. Boundaries determined by actual lateral-directional stability were considerably more restrictive than they were predicted to be. Although power-on wind tunnel test did indicate an effect of the rocket engine, tests of this nature are not conducted for most test programs."

Figure 3: Variations of XB-70-1 flight-based and predicted aileron yawing-moment control derivative with Mach number in hypothetical climbout profile. [B]

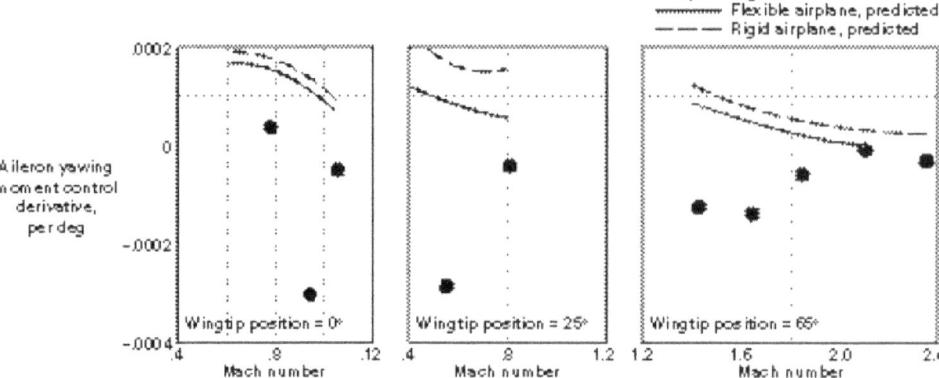

extremely significant since it caused a pilot induced oscillation (PIO) in flight.

We have seen some unusual power effects in each of the lifting bodies. These include longitudinal trim changes of as much as four degrees in angle of attack. Most recently, we have observed a loss of directional stability in the X-24B in the Mach number range from 0.9 to 1.0 and higher. This is illustrated in Figure 2. These power effects were not due to thrust misalignments [although some existed]. They were the result of rocket-plume induced flow separation over the aft fuselage, fins, and control surfaces. This phenomenon is apparently peculiar to lifting-body configurations or non-symmetrical shapes, since it had not been noted in earlier rocket aircraft or in missiles to any significant extent.[26]

Aerodynamic discrepancies are not limited to lifting-body configurations. We saw a reversal of sign in yaw due to aileron on the XB-70 as illustrated in Figure 3. Aileron characteristics

Figure 4: Calculated decrement/increment of lift-to-drag ratio resulting from the difference between predicted and measured base pressure coefficients in the XB-70. Only the lift-to-drag ratio increment in the shaded regions is used for range-increment calculations. [C]

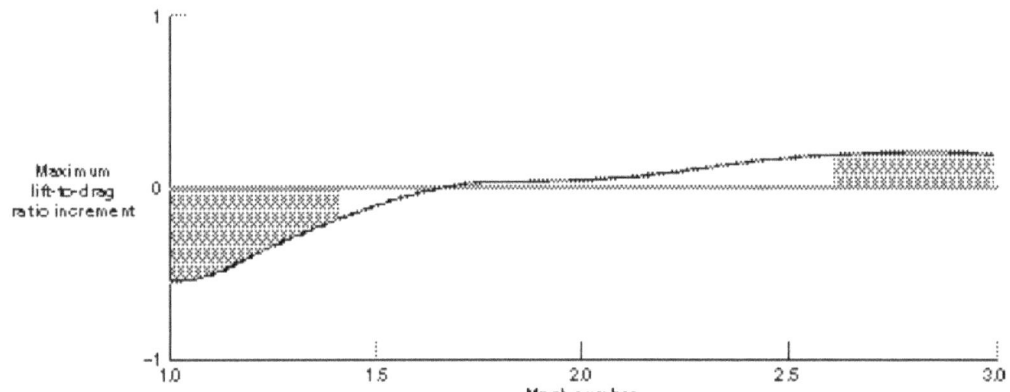

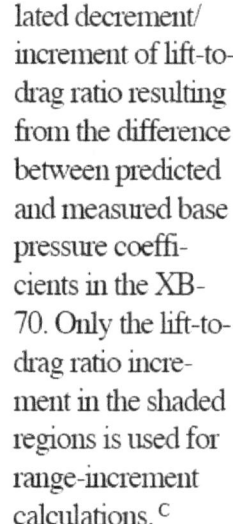

---

[26] Milt seems to be forgetting here that there were rocket-plume effects in the D-558-2 when any other cylinder of the XLR-8 rocket engine fired in a combination including the top cylinder. These effects were most severe at the highest Mach number tested—approximately Mach 1.6. The plume effects were small when only the two middle cylinders fired together in a horizontal plane. See Chester W. Wolowicz and Herman A. Rediess, "Effects of Jet Exhausts on Flight-Determined Stability Characteristics of the Douglas D-558-II Research Airplane" (Washington, DC: NACA RM H57G09, 1957), esp. pp. 16-17. There apparently were also plume effects on rockets such as the Saturn V.

[B] Taken from Chester H. Wolowicz, Larry W. Strutz, Glenn B. Gilyard, and Neil W. Matheny, "Preliminary Flight Evaluation of the Stability and Control Derivatives and Dynamic Characteristics of the Unaugmented XB-70-1 Airplane Including Comparisons with Predictions" (Washington, DC: NASA TN D-4578, 1968), p. 64. This may not have been the precise figure Milt had in mind, but it illustrates his point, showing that the predicted aileron yawing-moment control derivative was positive (proverse), whereas the flight-based values were negative (adverse) from a Mach number of about 0.90 through the supersonic range.

[C] Taken from Edwin J. Saltzman, Sheryll A. Goecke, and Chris Pembo, "Base Pressure Measurements on the XB-70 Airplane at Mach numbers from 0.4 to 3.0" (Washington, DC: NASA TM X-1612, 1968), p. 31. Again, this may not have been the exact figure Milt intended to use, but it makes his point. Notice that there was a favorable increment in lift-to-drag ratio at

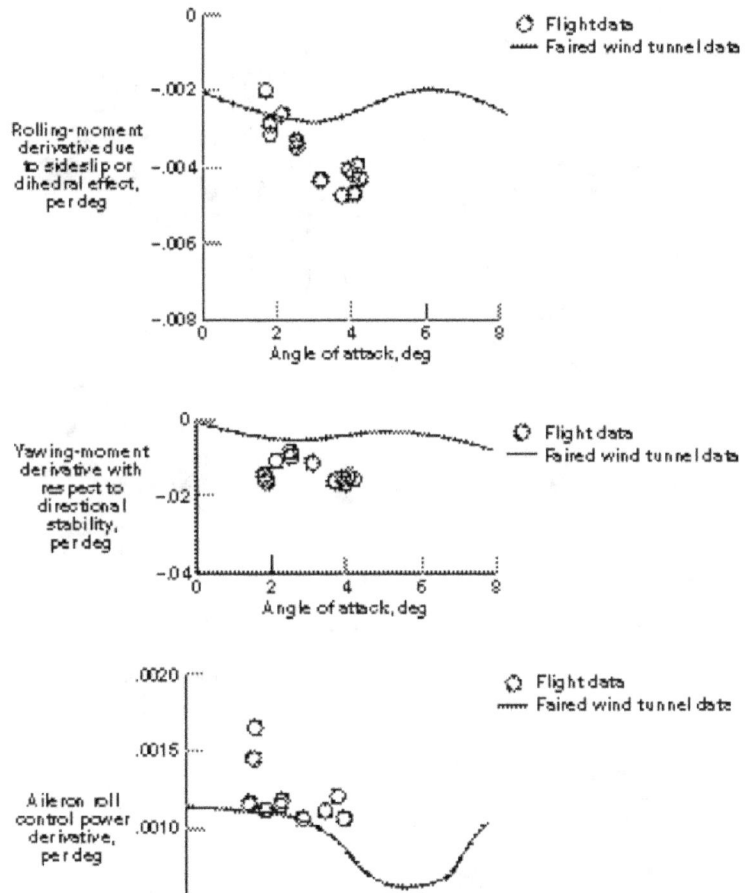

Figure 5: Lateral-directional derivatives as a function of angle of attack in the F-8 Supercritical Wing aircraft. [D]

couldn't be completely resolved even after the fact. It was suspected that the discrepancy was primarily due to flexibility or aeroelastic effects. After the flight program was completed, a new model was constructed and tested in an attempt to get better correlation between wind-tunnel and flight data. The best correlation that could be obtained—even knowing the answer in advance—was 10 percent on overall drag, and that still means a big error in overall flight range.

Aerodynamic discrepancies have not disappeared with time. During tests of the F-8 Supercritical Wing in the 1971-72 time frame, we saw numerous discrepancies between wind-tunnel and flight data even at the optimized design cruise condition of 0.99 Mach number. Figure 5 shows comparisons of wind-tunnel and flight data for some of the aerodynamic derivatives where significant discrepancies occurred. Admittedly, the design Mach number region is extremely hard to work in. Yet the 50- to 100-percent errors in such basic stability derivatives as the one for sideslip could hardly be considered acceptable accuracies. At other than design cruise condition, a large discrepancy was observed in aileron effectiveness and smaller but still significant discrepancies in the pitching-moment coefficients.

Here again, we had a master of the craft, [Richard] Whitcomb, conducting the wind-tunnel tests and analyzing the results before the fact. Admittedly, the airfoil concept was somewhat revolutionary; however, Whitcomb had essentially unlimited access to any wind-tunnel facility he needed and should therefore

of delta-wing aircraft have historically been hard to predict since the days of the XF-92, one of the first delta-wing aircraft. A discrepancy in aileron characteristics may not seem too significant, and yet an aircraft (a B-58) was lost during the flight-test program because of this particular error in prediction.

The B-70 drag discrepancy shown in Figure 4 resulted in a 50-percent reduction in predicted range. This is an excellent example of a discrepancy that

cruise speeds above Mach 2.5 but that at low supersonic speeds near Mach 1.2 there was the very unfavorable decrement Milt talks about. Thus, even though ground researchers had overestimated base drag at cruise speeds, their underestimate at low supersonic climbout speeds seriously reduced the aircraft's range.

[D] Taken from Neil W. Matheny and Donald H. Gatlin, "Flight Evaluation of the Transonic Stability and Control Characteristics of an Airplane Incorporating a Supercritical Wing" (Edwards, CA: NASA Technical Paper 1167, 1978), pp. 42, 43, 46. Once more, this may not be the precise illustration Milt intended to use, but it shows roughly the level of discrepancy between wind-tunnel and flight data that he had in mind and does so for some of the derivatives he mentions.

not have any good excuse other than the fact that the wind tunnels still have some obvious shortcomings.[27]

More recently, discrepancies in very basic stability characteristics have been observed in the newest and latest aircraft. The [Y]F-16 and [Y]F-17 showed a substantial difference between predicted and flight-measured longitudinal stability throughout a major portion of the usable angle-of-attack envelope.[28] The B-1 exhibited much more adverse yaw due to roll control on its first flight than had been predicted. This discrepancy showed up on a configuration and at a flight condition that should have been highly predictable.

In summation, we just haven't seen evidence to prove that wind-tunnel predictions are improving that much in accuracy or that we have gotten that much smarter in anticipating all the potential aerodynamic problems.[29]

Environmental Problems

To turn to the subject of environmental problems, I would like to review some that we experienced with the X-15. There have been numerous reports published and many papers given on the results of the X-15 flight program, but nothing has been published that summarized all the problems we had. We went back into the records to try to identify all the various problems. Before discussing them, however, we must recognize that the X-15 was quite an advanced aircraft for its time, except in terms of its configuration. This was pretty conventional as can be seen in Figure 6 with the possible exception of the upper and lower vertical tails, which were wedge-shaped. The aircraft had a unique structure for dealing with aerodynamic heating, and it featured many

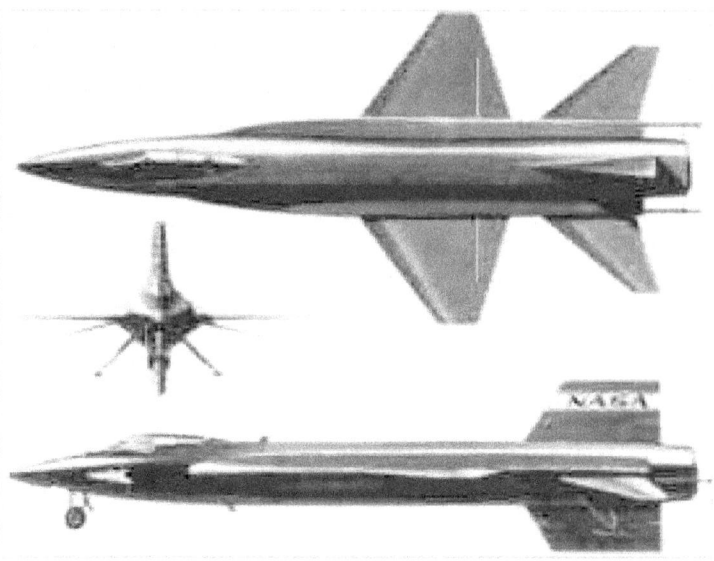

Three views of the X-15's original configuration, with which it achieved a maximum speed of Mach 6.06 and a maximum altitude of 354,200 ft. Its launch weight was 33,000 lb.; landing weight 14,700 lb. The lower half of its vertical tail had to jettisoned before landing, since as the little head-on view makes clear, it otherwise would have protruded below the landing gear when the latter was extended.

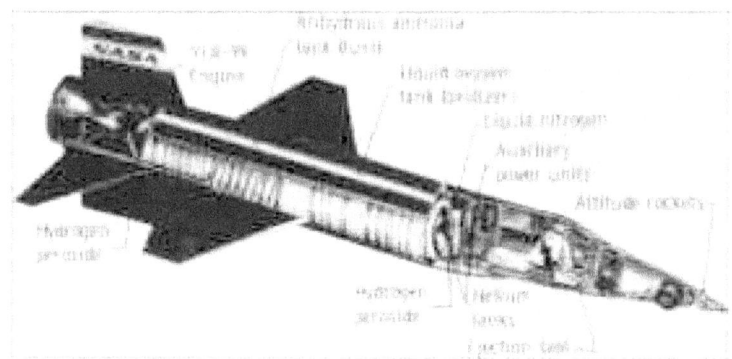

This cutaway drawing reveals the volume of tankage needed to give the X-15 its dazzling propulsion, it's pressurization, and it's attitude control in space. Liquid-oxygen capacity, 1003 gal.; anhydrous-ammonia capacity, 1445 gal.

Figure 6: Three-view and cutaway drawings of the X-15.[E] (See page 25)

---

[27] Note that Whitcomb discussed some of the preliminary differences between wind-tunnel and flight data in his "Comments on Wind-Tunnel-Flight Correlations for the F-8 Supercritical Wing Configuration," in *Supercritical Wing Technology: A Progress Report on Flight Evaluations* (Washington, DC: NASA SP-301, 1972), pp. 111-120, a report that was still classified when Milt was writing this document.

[28] The YF-16 and YF-17 were in a very close competition for an Air Force contract, which the YF-16 won in January 1975, and this led to the production F-16As—a fact that Milt could not have known at the time of his writing this document. The YF-17 later led to the Navy/Marine Corps F/A-18. See the Air Force Flight Test Center History Office's *Ad Inexplorata: The Evolution of Flight Testing at Edwards Air Force Base* (Edwards AFB, CA: AFFTC/HO, 1996), pp. 27-28.

[29] If Milt were writing today, he would no doubt add the results of Computational Fluid Dynamics (CFD) to his comments, since the results of CFD have also failed to anticipate many potential aerodynamic problems in vehicles that have used it as a design tool. On the other hand, many people would argue that wind-tunnel predictions have improved significantly, partly as a result of comparing previous predictions with the actual results of flight research, partly from other sources.

new systems that were required to fly to the limits of the flight envelope. These included the reaction control system, the inertial system, the LR-99 rocket engine with throttling, the skid [landing] gear, auxiliary power units, side-arm controller, the ball nose to provide air data, and the MH-96 Flight Control System (a rate command system with adaptive gain [that appeared only in the X-15 Number 3]).

As you might suspect and as will be discussed later, our major problems were with the systems rather than with configuration aerodynamics. In most areas, the aerodynamics were pretty much as predicted. There was good correlation between wind-tunnel and flight data throughout the entire Mach range.

The only significant difference was in base drag, which was 50 percent greater than predicted. Again, a characteristic historically hard to predict. The lift-to-drag ratio (L/D), however, was higher than predicted—4.5 as compared to 4.2—which indicates that there were compensating factors not evident in the wind-tunnel data. Ground-effect and gear-down L/D were also inaccurately predicted. One other important bit of data obtained during the X-15 flight program was aerodynamic heating data, which revealed that actual heat transfer rates were substantially lower than predicted by theory.[30]

Figure 7 addresses the X-15 program and some of the problems encountered. It

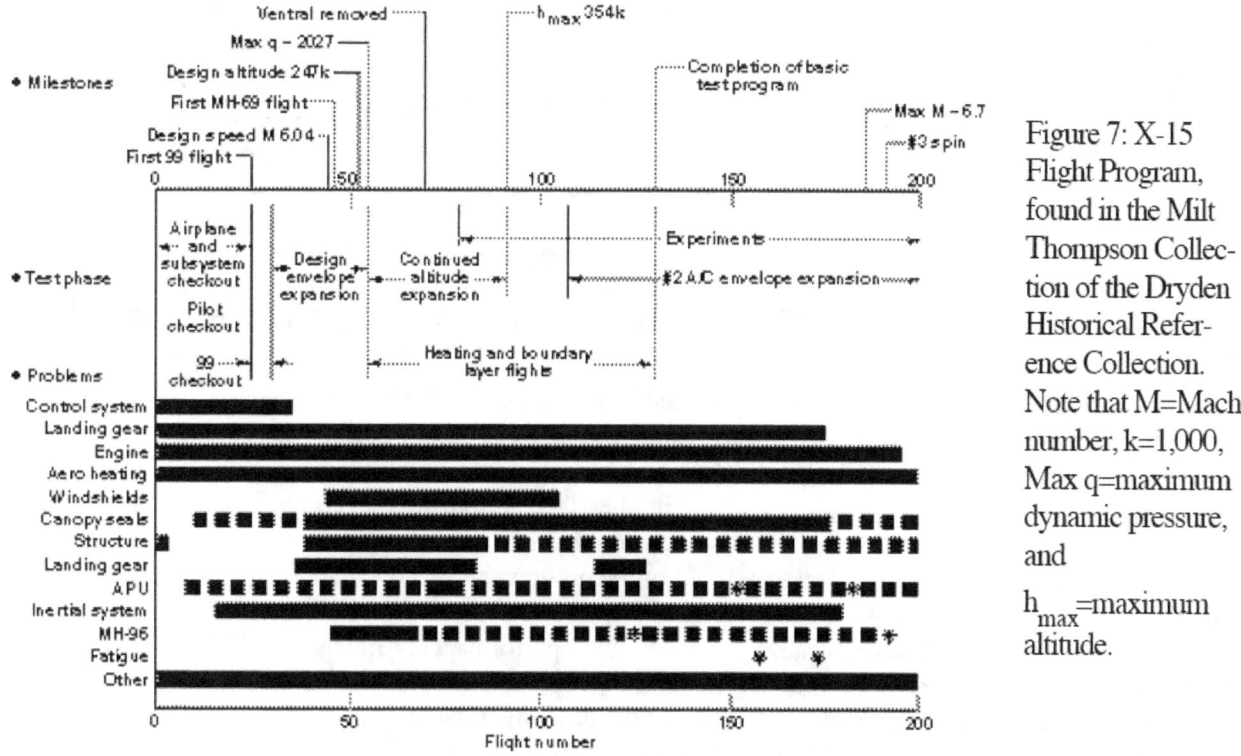

Figure 7: X-15 Flight Program, found in the Milt Thompson Collection of the Dryden Historical Reference Collection. Note that M=Mach number, k=1,000, Max q=maximum dynamic pressure, and h_{max}=maximum altitude.

E This was taken from Wendell H. Stillwell, *X-15 Research Results with a Selected Bibliography* (Washington, DC: NASA SP-60, 1965), p. 3.

[30] Another inaccurate prediction stemmed from the theoretical presumption that the boundary layer (the thin layer of air close to the surface of an aircraft) would be highly stable at hypersonic speeds because of heat flow away from it. This presumption fostered the belief that hypersonic aircraft would enjoy laminar (smooth) airflow over their surfaces. Because of this, many designers computed performance and heating for the hopeful case of laminar flow. At Mach 6, even wind-tunnel extrapolations indicated extensive laminar flow. However, flight data from the X-15 showed that only the leading edges of the airfoils exhibited laminar flow and that turbulent flow occurred over the entire fuselage. Small surface irregularities, which produced turbulent flow at transonic and super-sonic speeds, did so equally at speeds of Mach 6. Thus, designers had to abandon their hopeful expectations. On this matter, see John V. Becker, "The X-15 Program in Retrospect," 3rd Eugen Sänger Memorial Lecture, Bonn, Germany, Dec. 4-5, 1968, pp. 8-9; Albert L. Braslow, "Analysis of Boundary-Layer Transition on X-15-2 Research Airplane" (Washington, DC: NASA TN D-3487, 1966).

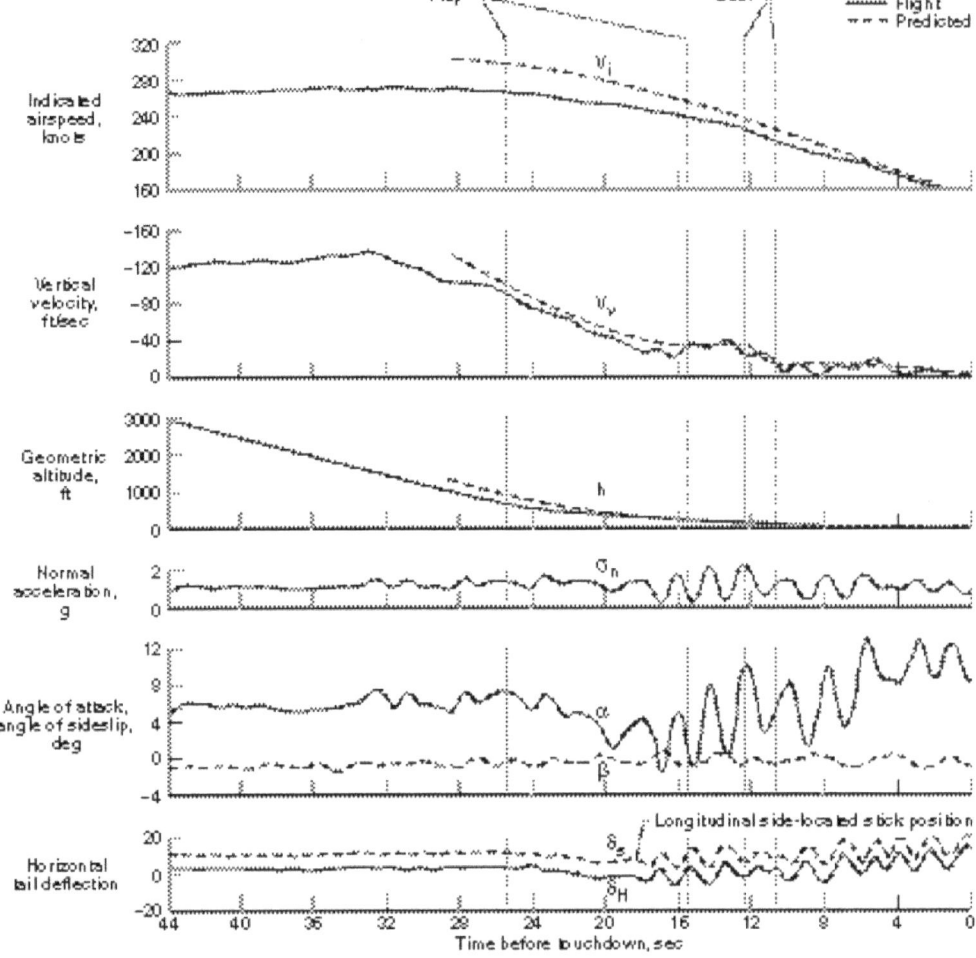

Figure 8: Time history of the flare and touchdown of X-15-1 on its first flight.[F]

indicates some milestones in the program and correlates them with the flight numbers. There was a total of 199 flights made with the three aircraft. In the middle of the figure, the various phases of the program are shown. On the bottom of the figure, some of the problem areas are listed. The bars and dashed lines indicate where the problems occurred during the program. The solid bar indicates continuing significant problems. The dashed lines indicate continuing minor problems, and the asterisks represent unique problems.

The Number Two aircraft was severely damaged on its 31st flight—the 74th X-15 flight of the program as a whole—and was subsequently rebuilt and modified to achieve higher performance.[31] It began flying again shortly after the halfway point in the program as shown in the test phase part of the figure.

[F] This was taken from Thomas W. Finch and Gene J. Matranga, "Launch, Low-Speed, and Landing Characteristics Determined from the First Flight of the North American X-15 Research Airplane" (Washington, DC: NASA TM-195, 1959), Fig. 13 on p. 26. This probably is the figure Milt had in mind to illustrate his point. On pp. 9-10, Finch and Matranga state:

> From [the] figure it is obvious that a severe pitching oscillation was induced near the end of the flap cycle. Reduced longitudinal trim was required as the flaps were being deflected, and the pilot added further airplane nose-down trim to avoid flaring too high. Apparently the oscillation became more severe because of the control input at about 18 seconds before touchdown. From this point, the pilot was not able to anticipate the oscillation accurately, which may have been aggravated by the fact that the control surface was rate-limited to 15° per second. . . . The transient in pitch covered an angle-of-attack range from -1° to 13°, with the amplitude as high as ±5°.

[31] On the 9 Nov. 1962 flight, Jack McKay could not get the XLR-99 engine to advance its throttle setting beyond 30 percent and had to make an emergency landing at Mud Lake under X-15 mission rules. He was unable to complete his jettison of propellants after

The envelope expansion to design speed, altitude, and dynamic pressure concluded with the 53rd flight—rather early in the program. It didn't take many flights to achieve these design conditions when one considers that 30 of the first flights were made with the interim engine—the LR-11 [two of which flew on each flight], which limited the maximum performance to about Mach 3. Once the LR-99 engine was available, the flight envelope was rapidly expanded—roughly half a Mach number at a time to the design speed of Mach 6, and 30,000 feet at a time to the design altitude of 250,000 feet. After achieving the design conditions, we began exploring the total flight envelope and continued to expand the altitude envelope, finally achieving an altitude of 354,[200] feet. We had the total impulse available to go even higher; however, the reentry was becoming somewhat critical. We also began exposing the aircraft to greater heat loads, going to high Mach numbers at lower and lower altitudes. We also began carrying piggyback experiments on the aircraft before the 80th flight and from the 130th flight on. That's essentially all that the X-15s were used for after that point since we had completed the basic aircraft flight-test program.[32]

Control-System Problems

The first major flight problem we had was with the control system, and this occurred on the first flight. The pilot got into a PIO during the landing flare. Very simply, the PIO was due to the limitation of the horizontal stabilizer to 15 degrees per second of surface rate and the pilot was asking for more than 15 degrees per second as illustrated in Figure 8. The

airplane was almost lost on the first flight as a result of this.

The PIO was a surprise because the simulation used to define the maximum control surface rate requirement did not adequately stimulate the pilot to get his own personal gain up. In the real environment on the first flight, his gain was way up. He was really flying the airplane. Our experience has verified that the pilot generally demands the maximum control surface rates for a given vehicle in the period just prior to touchdown, at least for unpowered landings.

In retrospect, this isn't hard to understand. Just prior to touchdown, the pilot is trying to control the flight path to within one-half a degree or so to make a good landing, five feet per second or less. An unpowered landing, in our opinion, is one of the most demanding tasks required of a pilot and a flight-control system. The problem is that you can't adequately simulate it. Visual simulators don't have the necessary resolution near the ground, and even sophisticated flight simulators such as variable stability aircraft can't seem to get the pilot's personal gain up sufficiently to thoroughly assess a potential PIO problem in landing. A PIO problem may not be evident until the first real unpowered landing is made. Even with a successful first landing one can't be sure the problem doesn't exist, since we have found that individual pilot gain varies considerably and another pilot may induce a PIO. The control system of the X-15 was modified after the first flight to increase the horizontal control surface rate from 15 degrees per second to 25 degrees per second.

---

shutting down the engine, and the excess weight caused him to be high on airspeed. He touched down at 296 miles per hour rather than the normal 230. The result subjected the main gear to both a rebound and a high aerodynamic load, causing the left landing gear to collapse, and eventually the aircraft flipped over on its back, injuring McKay and causing the Number Two aircraft to be rebuilt and modified. See Milton O. Thompson, *At the Edge of Space: The X-15 Flight Program* (Washington and London: Smithsonian Institution Press, 1992), pp. 227-230 and his further discussion of this flight below in this study.

[32] On the other hand, it could be argued that the hypersonic aircraft itself was the primary experiment from flight 1 to flight 199, even when it was carrying piggyback experiments.

The next major control-system problem didn't show up until the 23rd flight. The problem was a structural resonance problem wherein the Stability Augmentation System (SAS) was responding to the vibration of the structure on which the SAS box was mounted. This self-sustaining control-system problem almost shook the airplane apart during an entry from 169,000 feet. We subsequently added a notch filter to eliminate this problem. It surprised us because we did not conduct a structural resonance test. The X-15 SAS was one of the first high-gain, high-authority systems capable of responding to structural frequencies. Since that occurrence, we always conduct resonance tests of an aircraft with SAS on to look for such problems.

Structural Problems

We also had basic structural problems. On the 4th flight, one of the thrust chambers exploded during engine start, causing engine damage and a fire. The pilot shut down all the thrust chambers and jettisoned fuel before making an emergency landing on Rosamond Dry Lake. He was unable to jettison all the propellant because of the steep nose-down attitude. As a result, the aircraft broke behind the cockpit on nose-gear touchdown.

The aircraft designers had failed to anticipate the nose-down jettison problem. The aircraft were subsequently beefed up to handle this problem.

Landing-Gear Problems

Landing-gear problems plagued us throughout the X-15 flight program. The landing gear failed on the first landing. The landing gear was reworked and performed satisfactorily until the 74th flight. On that flight, after launch, the engine would only develop 30 percent thrust. The pilot was told to shut down the engine, jettison propellants, and make an emergency landing

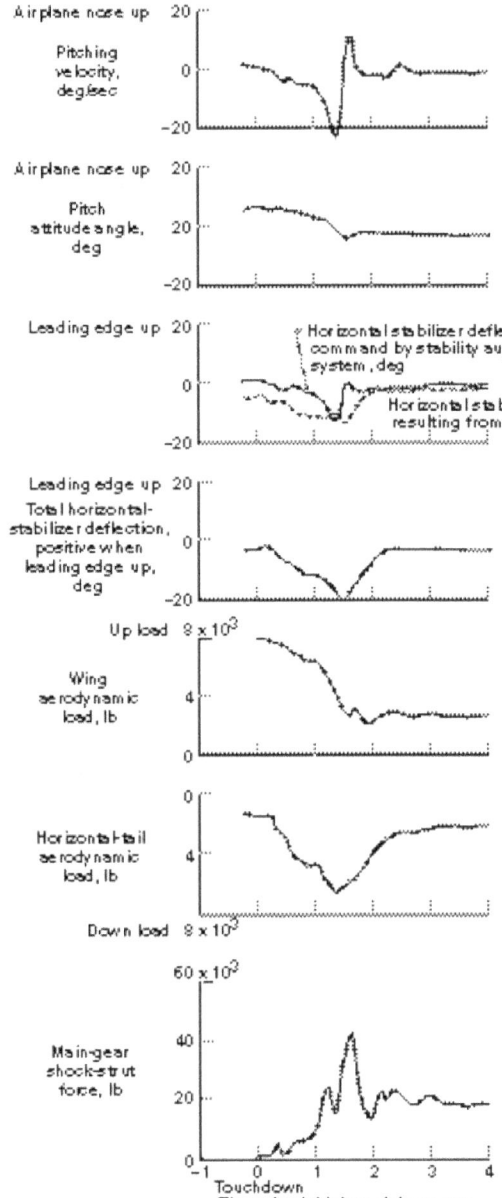

Figure 9: Typical X-15 landing using wing flaps. Nose-gear touchdown at a time interval between initial main-gear contact and nose-gear contact of 1.35 seconds (flight 1-30-51 on June 27, 1962). Taken from Richard B. Noll, Calvin R. Jarvis, Chris Pembo, and Betty J. Scott, "Aerodynamic and Control-System Contributions to the X-15 Airplane Landing Gear Loads" (Washington, DC: NASA TN D-2090, 1963), p. 26.

at the launch lake. Again, the pilot was unable to jettison all the propellants and, to compound the problem, the landing flaps did not extend when selected. The main gear failed shortly after touchdown and subsequently, the nose gear failed and the aircraft ended up on its back.

This gear failure resulted primarily from the high-speed touchdown due to the flap failure, and the high gross weight. Touchdown speed was almost 300 miles per hour. At main-gear touchdown, with skid-type gear, the nose tends to slam down rather rapidly.

Figure 10: A noteworthy scar from the X-15's first flight to Mach 6 was this cracked outer panel on the right side of the windshield. (NASA photo E-7508)

As the nose starts to pitch down, the SAS applies nose-up elevator to counteract the nose-down pitching moment. The airload at this high speed, resulting from the extreme deflection of the horizontal stabilizer located immediately above the main landing gear, plus the airload due to the negative three-point aircraft attitude, added to the normal rebound load from nose-gear impact, was sufficient to break the main landing gear.

A typical time history of loads on the main landing gear is shown in Figure 9. The air load problem due to SAS response was not fully appreciated in the initial design. A squat switch was later included to deactivate the SAS on main-gear touchdown. The squat switch worked quite well, but as the airplanes gained additional weight during the program due to added instrumentation, add-on experiments, and required modifications, additional fixes were required. The pilots were first asked to push forward on the stick at touchdown to relieve the air loads on the main landing gear. Later, a stick pusher and a third skid were added to prevent landing-gear failure. We were still having gear problems when the program ended after nearly 200 flights.

Aerodynamic Heating Problems

We had a number of problems associated with aerodynamic heating. They began showing up as we intentionally subjected the airplanes to high heating rates and temperatures. We had two windshields shatter, becoming completely opaque as shown in Figure 10, and four that cracked during flight. The shattering was due to failure of the glass itself at the high temperatures. An inappropriate choice of material was the cause. The cracking was due to distortion of the window frame at high temperatures. The support structure for the windshield glass was finally redesigned.

We had a problem with canopy seals. When the cabin was pressurized, the canopy leading edge deflected up just enough to allow the air to get to the canopy seal. At speeds above Mach 3, the air was hot enough to burn the seal, resulting in the loss of cabin pressure. The fix for this was to add a lip over the front of the canopy leading edge that prevented the air from impinging on the canopy seal.

We had a problem with local heating on the wing leading edges. Expansion gaps in the wing leading edge were designed to allow for the expansion due to aero[dynamic] heating. These gaps, however, triggered turbulent flow, which caused a hot spot directly behind the gaps. This caused the wing skin behind the gap to expand and pop the rivets holding the skin to the leading edge. Gap covers were added to eliminate this problem, but it persisted.

Aerodynamic heating also caused problems with the landing gear. The first problem was due to distortion of the nose-gear door. As the airplane got hot, the nose gear door tended to bow, opening a gap between the rear lip of the door and the fuselage skin behind the door. This allowed ram air to enter the nose gear compartment. The hot air cut through electrical wiring and tubing like an acetylene torch. The nose-gear door and its supporting structure were finally modified to eliminate this problem.

Another landing-gear problem due to aerodynamic heating resulted in the nose-gear-scoop door opening at a Mach number of 5.3. The nose-gear-scoop door was a small door designed to assist the extension of the nose gear aerodynamically. When it opened, it scooped ram air into the nose-gear compartment. At that speed, the air became hot enough to burn the tires off the nose wheels. The scoop door released due to distortion of the uplock linkage system under heating loads. The uplock system required a complete redesign.

The scoop door opened another time at Mach 4.3 on the modified Number Two X-15 because of a different problem with the uplock system. Again, the nose-wheel tires burned and when the pilot extended the landing gear just prior to touchdown, the nose gear extended very slowly. A nose-gear-up landing was barely averted because we had a sharp chase pilot who called the X-15 pilot to hold off until the nose gear was fully locked. The Number Two aircraft, which had been modified and rebuilt after a gear failure that resulted in a roll-over, thus had other gear problems attributed to aero [dynamic] heating.

The nose gear extended in flight at Mach 4.3 due to insufficient allowance for additional structural expansion in the landing-gear deployment cable system. The fuselage had been lengthened, but additional compensation for fuselage expansion had not been included in the landing-gear-cable release system. In another incident, the right-hand main landing gear deployed in flight at Mach 4.5 when the uplock hook broke as a result of the bowing of the main-landing-gear strut. The main landing gear on the modified Number Two aircraft had been lengthened to accommodate the supersonic combustion ramjet engine. The additional bending of the longer strut due to differential heating on the outer and inner portions of the strut had not been adequately compensated for, and the resulting deflection

in bending of the strut caused the uplock hook to fail in tension.

Auxiliary-Power-Unit (APU) Problems

We had APU problems during the early altitude-buildup phase of the program. No one had thought of pressurizing the APU gearbox cases. The lubricating oil was vaporizing at high altitude, and APUs were failing because of inadequate lubrication.

During a climbout on an altitude flight, the 184th flight,[33] one APU shut down because of an electrical transient that caused an electrical overload. When the first APU shut down, the electrical load shifted automatically to the second APU. The second APU should have accommodated the additional load, but because it was heavily loaded as a result of increased power demands over the years, it also shut down. The shutdown of both APUs resulted in a complete loss of hydraulic and electrical power as the aircraft was climbing through 100,000 feet. The aircraft virtually disappeared. The control room lost radar tracking, telemetry, and voice communications. The pilot lost the engine, all electrically driven instruments, and all control except for the manual reaction-control system operated by cables. He managed to get one APU restarted to regain hydraulic pressure for the aerodynamic flight-control system and successfully reentered the atmosphere from an estimated 160,000 feet altitude with no stability augmentation system and only a couple of instruments—a "g" meter and his barometric instruments.

MH-96 Problems

The Minneapolis-Honeywell flight-control system [MH-96] was fairly advanced for its time. It was a command-augmentation-type control system with adaptive gain scheduling and various

---

[33] This section is moved from a separate heading in the original typescript entitled "Other Problems." The original said this was the 154th flight, but as an anonymous reviewer of this publication correctly pointed out, it was the 184th flight, the 5 being an apparent typographical error.

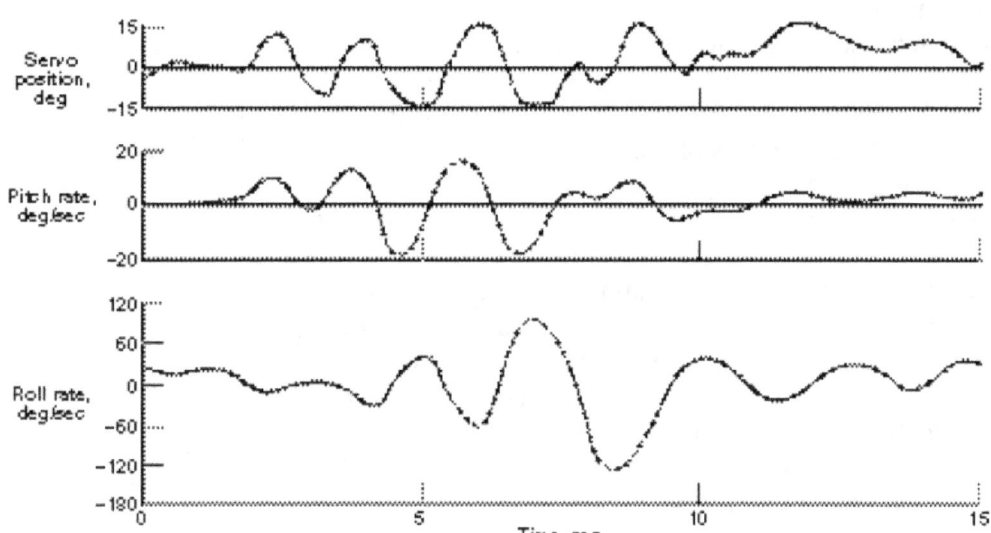

----- Estimated

Figure 11: X-15 No. 3 on flight 39 (3-39-62), 13 January 1965—the 125th flight of the X-15 program—in which the airplane became uncontrollable in pitch and roll for a short time.[G]

autopilot modes. It was installed in only the Number Three X-15 for evaluation.

At launch on the first flight, the stability augmentation portion of the system completely disengaged due to an electrical transient. The pilot managed to reengage it, but as you can imagine, it was quite a shock. Minor problems cropped up during the next 37 flights of the system. On the 39th flight [125th flight in the overall program], shown by an asterisk [in Figure 7], the system—you might say—went berserk. The horizontal stabilizers used for both pitch and roll control started limit cycling as a result of excessive gain through a deflection of ±10 degrees. During this limit cycling, which occurred at Mach 5.5,[34] the aircraft was essentially out of control in pitch and roll and was being oscillated by the motion of the horizontal stabilizers in both pitch and roll as shown in Figure 11.

It was quite a ride. Luckily the gain finally came down and the oscillation stopped. It took a while to find out the reason for this problem. It turned out that the system had fooled itself into believing it was at a flight condition where maximum system gain was required. The aircraft had been trimmed in a steady-state 4g [acceleration equal to four times the force of gravity] and the pilot was not making any inputs. Normal gain-reducing stimuli for the flight-control system were totally missing, and the system slowly drifted to maximum gain. When the pilot finally made a control input, the system went unstable. The electronics were saturated. This same problem later contributed to the structural breakup and loss of the number three airplane. We did not implement any specific fix for this problem after the first occurrence.

A cure for the problem was, however, discovered after some intensive simulator investigation, and we elected to continue to fly the system without modification. This particular problem could not be duplicated on the Iron Bird simulator after its initial occurrence. One of the reasons, we later found, that it could not be duplicated was that the hydraulic pumps supplying pressure to the Iron Bird

---

[G] The dashed line indicates that the roll and pitch rates exceeded the recorded limits and had to be estimated. Taken from Euclid C. Holleman, "Control Experiences of the X-15 Pertinent to Lifting Entry," in *Progress of the X-15 Research Airplane Program* (Washington, DC: NASA SP-90, 1965), p. 72. This appears to have been the figure Milt had in mind for this point in his narrative.

[34] It was actually Mach=5.35. Milt probably was giving just a ballpark figure.

simulator could not equal the output of the aircraft pumps. On the simulator, we could not physically rate-limit the control surfaces.

Another reason we could not duplicate the problem on the simulator was that we could not cause the system gain to hang up at maximum gain as it did in flight. It was finally concluded that the simulator gain would not hang up because of all the extraneous electrical noise in the simulation electronics. The electrical noise in the aircraft system was substantially lower. We finally managed to duplicate the problem on the simulator by physically pinning the system gain at its maximum. Once the problem was understood, the cure was obvious. The pilot had simply to reduce the gain manually and the limit cycle would cease. The pilot flying the Number Three aircraft on its final flight was aware of this potential problem and the required action should it develop. However, for unknown reasons, he did not take the proper action.

Fatigue Problems

Even though the total flight program included only 199 flights and only an average of 66 flights per airplane, we had what we considered a couple of fatigue problems. The 66-flight average in reality probably involved 200 to 300 system cycles when you include ground checkouts and aborts. The first fatigue problem was a rupture of the casing of the engine turbo-pump. The second was a rupture of a main bulkhead in one of the propellant tanks.

X-15 Program Results

With regard to environmental-type problems, the X-15 program has definitely convinced us of the desirability of a buildup-type test program when you have a lot of new systems that you are exposing to flight for the first time. If we had gone to the design speed and altitude on the first flight and had encountered all of the heating problems and the other subsystem problems simultaneously, we probably would have lost the aircraft. Regardless of all the problems we had, we did make a lot of successful flights.

The pilots, because they were designed into all systems, saved many missions and the aircraft itself on numerous occasions. Problems notwithstanding, Dr. [Hugh] Dryden [Director of the NACA and Deputy Administrator of NASA] referred to the X-15 flight program as the most successful research airplane program in the history of aircraft.

Control-System Problems in General

Flight-control systems are becoming more and more an integral element of new aircraft. Even now with the current generation of aircraft, the control-system design has in most instances been factored in to some extent before the configuration is finalized. It is therefore no longer practical to allow the aerodynamicist, the propulsion-systems engineer, and the structures people to design an aircraft as they have in the past since now flight-control technology has so much to offer. In a control-configured vehicle or active-controls-technology vehicle of the future, the control system will be factored into the initial design as early and extensively as the vehicle's aerodynamic, structural, performance, propulsion, stability-and-control, and other disciplines to achieve the optimum vehicle. The trend is obvious. Because of this trend, the flight-control system assumes much greater importance in the flight testing of a new vehicle. These new systems tend to use higher gains and authorities. They are thus more susceptible to such things as structural resonance, limit cycling, and surface-rate limiting.

Extensive ground testing of these systems is required to assure that the control system itself won't destroy the aircraft in flight. The structural-resonance and limit-cycle problems encountered in flight on the X-15 are excellent examples of the seriousness of the problem. It is essential, for example, that the flight-control system be active and monitored in all its various operating modes during ground-vibration testing. Other ground testing to define resonance and limit-cycle boundaries is also a mandatory requirement on any Flight Research Center flight vehicle.

Ideally, the control surfaces should be unloaded and then loaded during these tests to simulate the total hinge-moment environment that any particular surface can expect to be exposed to. For example, during the checkout of the M2-F2 for its first flight, some lead weights were placed on the upper flaps. The flight-control system was active at the time, and immediately, a control-surface oscillation of ±1 degree began. That particular problem resulted from a slight deflection of the power-actuator support structure under load. The support structure had to be beefed up to eliminate the problem. On another occasion, during limit-cycle testing of the HL-10 flight-control system, a limit cycle was intentionally induced at approximately 12 cycles per second. When the limit-cycle stimulation was terminated, the limit cycle continued. The stability-augmentation system was then disengaged in an attempt to stop the limit cycle, but it still persisted. Shutting off hydraulic power to the flight-control system finally stopped the limit cycle. That particular problem was a result of the servo-actuator and mechanical-feedback linkage dynamics.

The individual and combined effects of all other subsystem operations on the flight-control system should be examined as well as start-up transients of each sub-system. Fly-by-wire control systems will require additional pre-flight testing to ensure that no spurious inputs can get into the system. Lightning-strike problems have yet to be defined.

Even after performing all this ground testing, researchers should anticipate and make provisions for handling potential problems in flight. For high-gain, high-authority stability-augmentation systems, it is essential in our opinion that manual-gain-changing capability be provided the pilot during the flight-research program. The HL-10 limit-cycle problem discussed earlier, resulting as it did from higher-than-predicted control effectiveness, is a good example of the need for such capability in the cockpit. If through simulation, other potential handling-quality or control problems are identified, provisions should be made to vary the questionable parameter. The M2-F2 required a rudder-aileron interconnect to achieve adequate roll power throughout the flight envelope. The lateral-directional handling qualities were extremely sensi-tive to interconnect ratio as a function of angle of attack and dynamic pressure. Because of this and the fact that we could possibly have had some variations in predicted control effectiveness and thus, effective interconnect ratio, we provided the pilot an adjustable interconnect control.

Any potential PIO problem observed in simulation dictates consideration of a means of reducing stick gearing in conventional control systems or system gain in command-augmentation-type control systems. Stick-gearing-change provisions, however, are not easily provided in conventional control systems. We did not provide this capa-bility in the HL-10, although we knew from simulation that a control-sensitiv-ity problem might be encountered. As described earlier, the problem did arise, and only the skill of the pilot prevented a potential disaster. In command-augmentation or fly-by-wire control systems, effective gearing-change capability is relatively easy to imple-ment. Thus, there should be no good excuse for a serious or prolonged PIO problem in an aircraft with such a system. Yet they have occurred on first flights, indicating that someone didn't face up to the facts.

To reiterate, any PIO potential revealed through simulation should be taken quite seriously since we and others have been caught short so many times in the past. The absolute PIO potential is extremely hard to predict even with the most sophisticated moving-base or flight simulator. One of the major unknowns, as mentioned previously, is the pilot's own system gain. No simulator will stimulate the pilot to anything approaching a real-life first flight. On a first flight, the pilot's personal gain may be an order of magnitude greater than anything observed in simulation. To further complicate the problem, individual pilot gains vary substantially. One pilot, or even a series of pilots, may successfully fly a vehicle without a hint of a problem. Then, all of a sudden a pilot appears who can compete with the stability augmentation system in response. We have seen dramatic evidence of this.

Six research pilots successfully flew the M2-F1. The seventh pilot, an experienced test pilot, got into a divergent PIO immediately after takeoff on two successive attempts to fly the vehicle. The resultant slow rolls to a landing left even the most seasoned pilots in the world speechless. The same pilot, flying a different lifting-body vehicle, was actually able to compete, in response, with the stability augmentation systems at one cycle per second.

Command augmentation systems, mentioned earlier, are becoming quite popular. They are showing up in more and more of the newer aircraft. One might question whether they are really needed in some cases. Command augmentation systems generally do provide good control characteristics and are quite pleasant to fly. They do not, however, conform to MIL Specs [military specifications] in all respects, and they do have a number of insidious

characteristics. These systems tend to mask many of the cues the pilot normally relies on to give information or warning.

For example, a high-gain command-augmentation-type system tends to eliminate transients or trim changes due to gear or flap extension, or center-of-gravity changes. This may not seem significant, and yet these trim changes in the past have informed the pilot that the gear and/or flaps did indeed move when the appropriate lever was moved or that the center of gravity was not where you wanted it. A subtle thing—yet somewhat disconcerting when you don't have these cues.

These same control systems tend to provide invariant response throughout the flight envelope. This again would appear to be highly desirable; however, the variable response of the older control system warned the pilot of an approaching low-speed stall or overspeed just through feel alone. These new systems feel completely solid up to and sometimes over the brink of disaster, and thus artificial stall-warning systems are generally required. Speed stability is also generally lacking in an aircraft equipped with this type of system, and unless it is artificially provided, the pilot must continually monitor airspeed.

Normal dihedral effect and ground effect are also masked by systems such as these. A paper discussing many of these insidious characteristics was presented at an AGARD ([NATO] Advisory Group for Aerospace Research and Development) Flight Mechanics Panel meeting in 1967, sometime prior to the introduction of the F-111 into operational squadrons.[35] Yet at least one aircraft accident resulted from each of the insidious characteristics. Inadvertent high- and low-speed stalls

---

[35] Milt apparently was thinking of his paper with James R. Welsh, "Flight Test Experience with Adaptive Control Systems," presented at the AGARD Guidance and Control and Flight Mechanics Panels, Sept. 3-5, 1968, Oslo, Norway, which was a year later than he remembered and also a year later than when the first F-111As entered service with the U.S. Air Force, although only in limited numbers.

resulted in F-111 aircraft losses. A malfunction in the automatic fuel transfer system of one F-111 allowed the center of gravity to move well aft of the normal flight range and finally sufficiently far aft to cause the aircraft to diverge. The pilot of that aircraft was completely unaware of the problem until the aircraft diverged.

The inadvertent stall-spin problems in other aircraft equipped with these systems is still much in evidence. For some reason, the word is not getting around, and everyone has to learn the hard way. It's almost as bad as the old T-tail problem. The artificial cues being provided to warn the pilots of impending stall are in many instances completely inadequate, as are many of the stall inhibitor devices. There is still much more work to be done in these areas. These command aug[mentation] systems also in most cases have to be deactivated in a spin since they tend to apply improper spin recovery control.

Adaptive-gain flight-control systems potentially have problems maintaining programmed gains. Two different systems with which we are familiar have had histories of excessive gain problems and also too low a gain at times because of turbulence or other external stimuli that have not been adequately compensated for in the initial design.

Control Configured Vehicles (CCV) will without question be major contenders in the next generation of military and possibly commercial aircraft. The first step has already been taken in the YF-16. The concept is completely feasible; however, these control systems *must* have the predicted control moments and power and cannot be marginal on surface rates or hinge moments.

And, finally, automatic control systems are not infallible. Automatic flight-control systems are only as good as the people who designed them. If the designers have not anticipated all the possible situations or flight conditions that the pilot and aircraft can get into, trouble can result.

Admittedly, it is usually easier to make the desired or necessary changes through electronics, but it is still surprising to realize how many changes are made during a flight-test program on some of the newer systems. Twenty to fifty changes in the flight-control-system configuration are not uncommon in these newer systems. The changes required are in many instances minor changes or tweaking to optimize the system. We are aware, however, of some major changes that were required such as gearing changes as high as fifty percent of the original value. The fact that major changes such as these are required is quite disturbing since these aircraft are not exploring new flight regions. Thus, the predicted aerodynamic data should be good as far as basic stability and control derivatives are concerned, and these are the primary requirements for the design of a flight-control system. The reason for such drastic changes being required is therefore not clear. Somewhere, somehow, something is being overlooked or not being considered in the design process.

Finally, the primary message on these new control systems is to shake, rattle, and roll them thoroughly before flight and then expect problems in flight and provide the necessary system-adjustment capability to alleviate the problem if it does occur.[36]

First Flight Preparation

The aerospace industry has had much more first-flight experience than either NACA-NASA or the military. It has been only recently that the government has been directly involved in preparing a

---

[36] Despite the fact that this section, like the rest of this document, was written about 1973 or 1974, much of what Milt says is still applicable, although in many cases pilots have adjusted their flying styles to adapt to the circumstances imposed by control systems, such as those in trim.

vehicle for and participating in first flights, and we may therefore not be the most qualified to define the best approach. The current problem, however, is that very few new aircraft have been designed and flown in the past fifteen years, and therefore even industry has had little recent experience. Many aerospace companies were essentially without a real flight-test organization during the lean years, and some did not have company test pilots. Thus, with the renewal of aerospace activity, many companies had to put a flight-test team together from scratch. That can spell trouble.

We at the NASA FRC have been fortunate in being in a position to actually conduct a number of first flights over the past ten years. We made first flights on all of the lifting bodies (the M2-F1, M2-F2, M2-F3, HL-10, the X-24A, and the X-24B) as well as the F-8 Supercritical Wing, the F-8 Digital Fly-By-Wire, the F-111 Transonic Aircraft Technology, and most recently the F-15 Remotely Piloted Research Vehicle. We have gained some appreciation for all of the concerns that go along with making a first flight and have developed our own procedures and ground rules for qualifying a vehicle for flight. These are by no means all-inclusive, since in many disciplines we depend on the designer and builder for the necessary confidence to proceed.

For example, we depend heavily on the contractor to provide an adequate structure and functional systems. We have in many cases done the conceptual design but have never attempted to do the detail design since we are not designers—a fact that others in government don't always admit. We do monitor and analyze the design, do proof loading of critical portions of the structure, and do functional testing of all the subsystems once a vehicle is delivered. We also do all the other normal pre-flight things such as taxi tests and all-up rehearsals of the first flight with all systems operating and with all personnel at their appropriate stations.

One of the most important things we do is simulation. We have learned from experience that extensive simulation is the key to success in flying a new configuration or vehicle. We analyze all the wind-tunnel data and then start with the best fairing of all the data. Once we complete that evaluation, we then begin looking at the worst cases. We intentionally vary each and every derivative over a wide range to determine the sensitivity of the vehicle's flying qualities to that particular derivative. The range of variability we investigate is much broader than the scatter of the wind-tunnel data. From experience, we expect—or I should say we are not surprised by—discrepancies of 25 to 30 percent in predicted derivatives. In practice, we vary them as much as 100 to 200 percent. In the case of dynamic or rotary derivatives, which are hard to measure both in the wind tunnel and in flight, we may vary them even more. Based on experience, we even vary combinations of various derivatives to look for the worst possible cases. Any potential handling-quality problems exposed in this type of investigation are thoroughly explored to determine possible fixes, recovery techniques, and/or, if necessary, ways to avoid the problem area. The low-angle-of-attack PIO problem identified during early simulations of the M2-F2 was a classic example. We spent many hours evaluating the problem and determined that we had two effective recovery techniques. One was to reduce the rudder-aileron interconnect ratio and the other was simply to pull up and increase angle of attack. We could not easily avoid the area since we had to go to low angles of attack to pick up the necessary airspeed for landing flare. We did, however, have landing rockets as a backup in case we did have to increase angle of attack from the desired pre-flare condition. In flight, both of these recovery techniques were validated by necessity. The value of this type of simulator investigation cannot be over-emphasized. We also feel that it is essential that the pilot participate extensively in the engineering simulation. We at FRC rely heavily on the pilot for a successful flight program. The importance of the pilot is critical, particularly in simulation, the design or evaluation of the control system, and first-flight preparation.

Pilots generally have a much broader and more objective view of the overall picture [than do other participants in flight research]. We are fortunate to have exceptionally well qualified and experienced pilots, all of whom have made first flights. They are all well suited to serve as either a project pilot on a new vehicle or a member of the Flight Readiness Review Board, and in this manner, we get double the input. A pilot with first-flight experience is invaluable. Unfortunately, there aren't too many active pilots who have first-flight experience because of the limited number of new aircraft that have been produced in the last fifteen years. Following the extensive engineering simulation, we proceed into the procedural and flight-planning simulation phases. In these phases, we develop the first-flight plan and then look at every imaginable failure, malfunction, or emergency. In developing the control system and preparing for the first flight of the M2-F2, we spent at least a year on the simulator, and I as pilot averaged two to three hours a day in the simulator cockpit.

We have not normally used anything other than a fixed-base simulator; however, we were sufficiently concerned about the low angle-of-attack PIO problem in the M2-F2 to also investigate it in-flight with a variable-stability aircraft. We have resorted to variable-stability aircraft and more sophisticated simulators on a few other occasions; however, the simple fixed-base simulator has generally been adequate. We thoroughly exercise the flight-control system to establish limit-cycle boundaries and ensure that we are free of structural resonance problems as discussed earlier.

Another very important part of our pre-flight preparation is a Flight Readiness Review (FRR). An FRR team is designated at least six months prior to a scheduled first flight. The team is generally composed of members of each of the disciplines involved (aerodynamics, stability and control, performance, etc.) as well as subsystem experts, instrumentation experts, and a pilot. The chairman of the FRR team is at least a senior division-director-level individual with a broad test background. The FRR team members are not associated in any way with the project team and act as devil's advocates. The FRR team has unlimited access to any data, can monitor any tests, question any project team member, and make any recommendations on pre-flight preparation. In essence, it has *carte blanche* to examine the program. The FRR reports directly to the Center Director, and prior to flight, it submits an oral and written report. The FRR is an extremely effective means of ensuring a safe first flight.

First Flight and Envelope Expansion

Our general philosophy on first flights is that once the aircraft is airborne, we immediately begin worrying about how to get it back on the ground again safely. Data maneuvers are of secondary importance. The main area of interest during the first flight involves the approach and landing tasks. Various potential failure or backup control modes are evaluated in the approach-and-landing configuration, as are other possible subsystem malfunctions. This emphasis on approach and landing during the first flight is easily justified, since on each and every succeeding flight an approach and landing must be made. Subsystem malfunctions will ultimately occur during the test program, and it is thus wise to assess these potential malfunctions in a controlled manner as early as possible.

Our philosophy on envelope expansion is not unique. We select the most benign

Mach number expansion corridor and concentrate first on validating stability, control, and handling-quality characteristics. We update the simulator following each flight [to incorporate what we learned on that flight that we did not know before]. Once the Mach number envelope has been explored, the remainder of the flight envelope is somewhat systematically expanded in terms of angle of attack, dynamic pressure, etc. In the case of the X-15, we began expanding the altitude and angle-of-attack envelope about halfway through the Mach envelope expansion. Once the design Mach and altitude had been achieved, we continued to expand the altitude envelope and simultaneously began expanding the dynamic pressure and aerodynamic heating boundaries. As previously indicated, this type of envelope expansion is typical of the general approach used throughout the aircraft industry and is a time-proven way to test aircraft.

In such an incremental buildup of flight research, we can, so to speak, poke our noses into a new area and if we encounter a problem, we can immediately back out of that area and into a known safe-flight region where we have flown before. When we do encounter a problem in flight, we come back down, analyze the data, update the simulator, and then try to determine a fix for the problem before we again probe into the problem area. The longitudinal-sensitivity problem we observed on the first flight of the HL-10 is a good example. Flight data confirmed that we had more control effectiveness than we anticipated. We made a change in the control-system gearing before the next flight and eliminated the problem.

During the buildup test program in the X-15, we were fortunate to encounter our environmental problems one by one. We burned the canopy seal at Mach 3.3, well below the design Mach number of 6.0. We encountered the

nose-gear-door problem at Mach 5. We saw the first indications of the wing-leading-edge and windshield problems at 5.2 Mach number. The first nose-gear scoop door opened at 5.5 Mach number. If we had gone to Mach 6 on the first flight, we would probably have had all of these things happen within seconds of one another. Also, each problem would have been more severe than we actually experienced because we would have had more exposure time.

Remotely Piloted Research Vehicles (RPRVs)

The Flight Research Center has developed a remotely piloted research technique that was first applied in the testing of an advanced lifting entry configuration, the Hyper III. The technique illustrated in Figure 12 includes basically a ground cockpit, an uplink for command control signals, and a telemetered downlink that closes the control loop through the pilot's instruments and controls. The cockpit has all the conventional instruments and controls normally found in an aircraft, and the pilot thus has complete instrument-flight-rules flight capability. In addition, a forward-looking television mounted in the flight model provides the ground pilot an out-the-window view for additional reference. A high-speed computer is included in the control loop to provide or exactly duplicate any flight control system augmentation or automatic control mode. This allows for a relatively simple and inexpensive on-board control system.

This technique has recently been applied to spin testing. A 3/8th-scale model of the F-15 has been tested throughout the achievable angle-of-attack range and intentionally spun using several different control modes. As of this time, the model had not been departed or spun using the primary control mode with the operational stall-

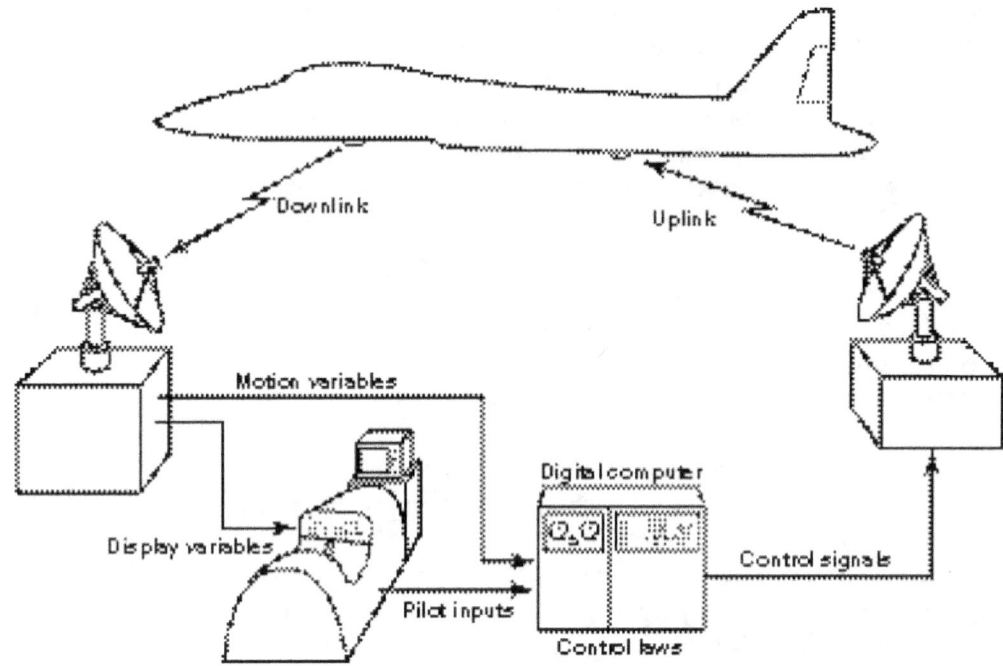

Figure 12: The remotely augmented vehicle concept.[H]

inhibitor system in the loop. We are still trying, however.

The flight program has, in our opinion, been an outstanding success even though we have damaged and finally lost the first flight vehicle. We anticipated losses and originally had three vehicles constructed to ensure completing the planned flight program. In the fourteen flights that have been accomplished,[37] we have thoroughly documented the stability and control characteristics of the model from plus 40 degrees to minus 20 degrees angle of attack. We have spun the vehicle upright and inverted and departed the vehicle at several different g-levels. We have validated two different spin modes predicted in the Langley spin studies and confirmed the proposed recovery techniques. We have not as yet compared model data with full-scale airplane data, since the full-scale flight data has so far been unobtainable. The 3/8-scale-model data has so far compared quite favorably with wind-tunnel predictions. The simulator developed during the flight program is probably the first good spin simulator ever implemented. Flight results have confirmed the validity of the simulator, and the simulator can and has been used to investigate and develop new spin entry techniques.[38]

[H] This was taken from Dwain A. Deets and John W. Edwards, "A Remotely Augmented Vehicle Approach to Flight Testing RPV Control Systems," paper presented at the AIAA RPV Technology Symposium, Tucson, AZ, 12-14 Nov. 1974 (also published at Edwards, CA, as NASA TM X-56029, Nov. 1974), p. 17.

[37] Readers who skipped the background section may like to know that the vehicle completed 27 flights by 1975, 53 by mid-July 1981.

[38] Kenneth W. Iliff, "Stall/Spin Flight Results for the Remotely Piloted Spin Research Vehicle," paper presented at the AIAA Atmospheric Flight Mechanics Conference, Danvers, MA, 11-13 Aug. 1980 (AIAA Paper 80-1563) gave the results of flight research with the 3/8-scale F-15, later redesignated the Spin Research Vehicle, most of the way through its flights. Among other findings it reported were: "the basic airplane configuration was found to be departure and spin resistant. When control authority was increased, the model could be spun using several techniques developed with the simulator." Also: "The acquisition of high quality steady spin data for this vehicle was made possible by the remotely piloted technique."

Much more remains to be done in the F-15 spin program, but whether it will be completed is questionable because of other higher-priority flight-program commitments. Higher-Mach-number departures and spins should be investigated. The model should be modified to provide more representative inertias, since the present inertias are higher than that required for dynamic scaling. The inertias are properly ratioed, however. Additional stability-and-control derivatives should be obtained during the actual spin. Wind-tunnel data is lacking at the extreme angle of attack investigated in flight. And finally, wind-tunnel, small scale-model, 3/8th-scale model, and full-scale data should be compared to completely validate the RPRV technique. All of this, we feel, is essential to ensure that in the future accurate simulations can completely predict departure and spin techniques as well as recovery techniques.

The current FRC position on RPRVs is that they can be used effectively to provide meaningful and accurate data. They can be cost effective, and they can save time and potentially even reduce the amount of full-scale testing required. The RPRV technique is extremely attractive for high-risk-type testing such as spin testing, testing of new structural concepts, testing of flutter-suppression systems, etc. A number of such programs have already been proposed, and we anticipate many more to materialize. Our problem now is to maintain some reasonable balance between unpiloted and piloted flight programs.[39]

There is a definite role for RPRV-type testing based on what we now know. The RPRV approach is, however, by no means a panacea for flight testing. RPRV tests may still have to be supplemented by piloted testing, and thus it may not always be most cost effective overall to go the RPRV route. We still have a lot to do in developing the technique and reducing the potential loss rate. It will also be some time before the reliability of the on-board pilot can be reproduced in RPRVs.

Flight-Test Errors

The remarkable safety record mentioned in the introduction does not imply that we have been without fault. The M2-F2 landing accident is a good example of poor judgment on our part. The M2-F2 at best had marginal lateral-directional handling characteristics. The pilot initiated a serious PIO inadvertently on the first flight of the vehicle. Another PIO occurred on a later flight during a data-gathering maneuver involving another experienced research pilot. Following the second occurrence, we should have quit flying the aircraft and gone back to the wind tunnel to look for a fix. We chose instead to continue flying the vehicle without modification. A third pilot, who had previously flown the vehicle, encountered a PIO on final approach on the [six]teenth flight. He successfully recovered, but as a result of the PIO, he was forced to attempt a landing on an unmarked portion of the lakebed.

Depth perception on the lakebed is extremely poor without known reference marks. To further complicate the problem, a rescue helicopter was operating in the immediate area of the modified landing site. This distracted the pilot because of a

---

[39] Following the flight research with the 3/8th-scale F-15/Spin Research Vehicle, the Center flew research programs with the Mini-Sniffer, the Oblique-Wing Research Aircraft, Drones for Aerodynamic and Structural Testing, and the Highly Maneuverable Aircraft Technology , among other remotely piloted vehicles. This range of vehicles showed that although Milt's comments in 1973 or 1974 were accurate as far as they went, sometimes—as he suggests below—sub-scale vehicles could be more expensive and time-consuming than full-scale programs because of the need to develop miniature systems to accommodate the smaller spaces in the vehicles. In other cases, however, RPRV operations could be cost effective, especially if flights were planned for high data output.

concern about a possible collision with the helicopter. The chase aircraft were also forced out of position during the PIO and were not in position to call out height above the lakebed. The result was a gear-up landing. The pilot suffered the permanent loss of vision in one eye. The vehicle was rebuilt with a center vertical fin after additional wind-tunnel tests predicted a significant improvement in flying characteristics. The flight research with the modified vehicle proved to be completely uneventful, since the flying qualities were quite good, as predicted.

We have had a number of problems in the past during towing operations involving unconventional flight vehicles. Some of these were due to our overall inexperience with aerial towing. We reinvented many of the problems well known to glider and sailplane people even though we had an experienced sailplane pilot as Center Director. We also used poor judgement when we decided because of cost to do our own towing. None of our pilots had any real towplane experience and only a couple had ever been exposed in any way to towing operations. As one might expect, we had several problems and one serious accident, luckily without any pilot injury.

The events leading up to the accident started with the acquisition of an L-19, which we modified for towing the paraglider. The pilot selected to do the towing had never flown an L-19, and I was elected to check him out. Because of various schedules, only one day was available for checkout. I rode through two flights with the newly selected tow pilot and undoubtedly overrode the controls, particularly the rudder, during landing approaches. I was then called away to a meeting and decided on the spot to let him take it alone. On his first landing, he ground looped and severely bent one main-gear strut.[40]

The strut was replaced and a towing flight was scheduled the following day. On the morning of the scheduled flight, the pilot who had been checked out was rescheduled for a higher priority flight. Another pilot was selected to do the towing, and I gave him a quick checkout in the L-19 prior to the actual towing flight. After takeoff and upon reaching the edge of the lakebed, the tow-plane pilot, as instructed, began a turn to stay close to the lakeshore in case of a tow-line failure. His rate of turn was excessive, and within seconds, the tow-line was hanging slack between us. Since we were only 300 feet or so high, the only recourse was to release and attempt a landing in the sagebrush. The vehicle was extensively damaged in the landing attempt. Following this accident, we reverted back to using professional tow pilots. We finally gave up towing altogether after two hair-raising incidents while towing the M2-F1.

The loss of the Number Three X-15 could be attributed to some extent to a faulty experiment that we developed and flew on the aircraft. The total experiment did not undergo a complete environmental check, although a component of the experiment, the drive motor that caused a problem, had successfully passed all environmental checks and had been used in other piggy-back experiments carried on the aircraft. The motor began arcing at approximately 80,000 feet altitude on the way up to a planned maximum altitude of 250,000 feet. The experiment was supposedly isolated from all primary aircraft systems, and yet it caused faulty guidance-computer and flight-control-system operation. This is another potential problem to be assessed with command augmentation and fly-by-wire control systems. The faulty experiment cannot be completely

---

[40] Milt added here, "We found out later that that particular pilot had never flown a tailwheel airplane." The pilot in question wrote beside these words, "Not true. I was the pilot involved. I flew the T-6 210 hours in pilot training."

excused of partial blame for the ultimate loss of the aircraft.[41]

Another example of questionable judgment on our part involved the maximum speed flight of the Number Two X-15 to a Mach number of [almost] 7. The flight was made to demonstrate the capability of the X-15 to carry a supersonic combustion ramjet (scramjet) engine to Mach numbers approaching 8. Figure 13 shows the X-15A-2 with a dummy ramjet on the lower stub ventral [, eyelid, drop tanks, and ablative coating[42] for what turned out to be the Mach 6.7 flight]. In building up to fly this combined configuration, we first made a flight with empty tanks to demonstrate tank jettison capability.

---

[41] Here Milt inserted a comment, "Show time history of X-15 #3." In lieu thereof, perhaps the section of the accident report for the aircraft quoted in his *At the Edge of Space*, p. 263, will better indicate the problems that caused the pilot, Michael Adams, to lose his life in the accident:

> The only unusual problem during the ascent portion of the flight was an electrical disturbance that started at an altitude of 90,000 feet and that effected [*sic*] the telemetered signal, the altitude and velocity computer associated with the inertial platform and the reaction controls that operate automatically in conjunction with the MH-96 adaptive control system. Although the pilot always had adequate displays and backup controls, the condition created a distraction and degraded the normal controls. As the aircraft approached the peak altitude of 266,000 feet, it began a slow turn to the right at a rate of about 0.5 degrees per second. The rate was checked by the MH-96 system which operated normally for a brief period so that at peak altitude, the aircraft was 15 degrees off heading. Then the pilot, apparently mistaking a roll indicator for a sideslip (heading) indicator[,] drove the airplane further off in heading by using the manual reaction controls. Thus the aircraft was turned 90 degrees to the flight path as the aerodynamic forces became significant with decreasing altitude. The aircraft continued to veer and entered what appeared to be a classical spin at an altitude of about 230,000 feet and a Mach number of about 5.0.

> Some combination of pilot action, the stability augmentation system, and the inherent aircraft stability caused the aircraft to recover from the spin at an altitude of about 120,000 feet and a Mach number of about 4.7. As the aircraft recovered from the spin, however, a control system oscillation developed and quickly became self-sustaining. At this time the airplane was descending at a rate of about 160,000 feet per minute and dynamic pressure was increasing at nearly 100 pounds per square foot each second. There was a corresponding rapid increase in the *g* forces associated with the oscillation, and structural limits were exceeded. The airplane broke into many pieces while still at high altitude probably in excess of 60,000 feet, and fell to earth northeast of Johannesburg, California.

> The pilot, probably incapacitated by the high *g* forces[,] did not escape from the cockpit and was killed on ground impact. The accident board concluded that the accident was precipitated when the pilot allowed the aircraft to deviate in heading and subsequently drove it to such an extreme deviation that there was a complete loss of control. The board believes that these pilot actions were the result of some combination of display misinterpretation, distraction, and possible vertigo. The board further concludes that the destruction of the aircraft was the result of a sustained control system oscillation driven by the MH-96 adaptive control system that caused the divergent aircraft oscillations and aerodynamic loads in excess of the structural limits. The electronic disturbance was attributed to the use in one of the scientific experiments of a motor that was unsuited to very high altitude environments.

Milt said he did not believe that there was any pilot error and that he thought the accident board agreed with him. He did think that the vertigo contributed to the accident (pp. 263-264).

[42] When the X-15A-2 was rebuilt from the Number Two X-15 following its landing accident, it gained an elongated fuselage and a small internal tank within the plane. Because the ablative coating put on the aircraft to protect it from severe heating on the higher-speed flights would char and let off residue at very high velocities, North American had placed an eyelid above the left cockpit window. The pilot would keep it closed until just before the approach and landing, using the right window for visibility during launch and most of the remainder of the flight. Above Mach 6, however, the residue coming from the charred ablator would cover the exposed window and restrict visibility. Hence the need to open the eyelid for approach and landing.

Figure 13: Photo showing X-15A-2 with ablative coating, drop tanks, and dummy ramjet. (NASA photo ECN 1889)

Next we made a flight with full tanks to demonstrate proper fuel and lox [liquid oxygen] transfer. The flight was aborted shortly after launch when there was no indication of flow from one of the tanks. The next full-tank flight was successful to a Mach number of 6.3. The next flight was with the eyelid and dummy ramjet to Mach 4.75.

At this point in the program there was a strong desire to put it all together and go for an all-out flight. The argument was that we had adequately demon- strated each of the configuration changes. We compromised for another flight with the ablator, eyelid and dummy ramjet. This flight raised the speed to Mach 5. We examined the airplane after that flight and saw some indications of localized charring but nothing that we considered significant. We simply made local repairs to the ablator and put it all together for a flight to Mach 7.0. Figure 14 shows [some of] the results.

A shock wave off the dummy ramjet interacting with the boundary layer caused severe localized heating that burned off all the ablator and burned through the basic Inconel ventral fin

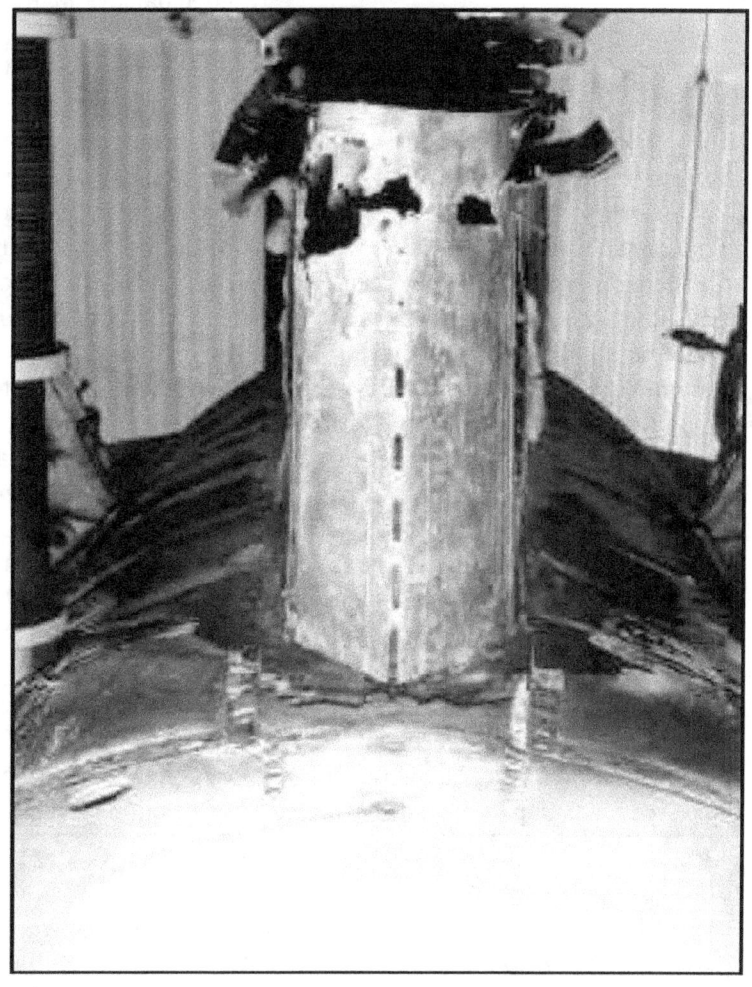

Figure 14: Result of severe heating from the Mach 6.7 flight in the X-15A-2. (NASA photo E-17525)

structure. We almost lost the airplane. We took too big a step. We essentially went from Mach 5 to almost Mach 7 in one step with the dummy ramjet. The moral is that even though we supposedly checked out each individual configuration, we should have put them all together and again worked up incrementally in Mach number. We would thus have appreciated the significance of the shock-impingement heating problem. As mentioned in the introduction, we have been in the flight-research business over twenty-seven years. Many of the people who worked on the X-1s are still here, and yet we still occasionally get caught short. We seldom get caught on the same problem, but it seems that we never run out of new problems.

Conclusions

Irrespective of the fact that our new generation of aircraft (F-14, F-15, F-16, F-17 [precursor of the F/A-18], and B-1) is not probing new frontiers, we are still seeing discrepancies between wind-tunnel and flight data as mentioned earlier. The X-1 achieved a Mach number of 2.5 over twenty years ago and we are still operating within that Mach region with most of our new aircraft. The aerodynamic discrepancies or problems we are currently seeing are not as dramatic as the loss of directional control and consequent tumbling that Chuck Yeager encountered in the X-1.[43]

We don't expect surprises such as roll coupling or aileron reversal, but we are impressed for example with the unexpected high-angle-of-attack capability of the F-14. This was not anticipated or at least not advertised. We still are occasionally disappointed in actual airplane performance. We still see substantial discrepancies between predicted and actual basic stability derivatives on occasion, which means we haven't improved our predictive capabilities or techniques substantially in the last twenty years regardless of any new or improved ground facilities.

The many discrepancies between wind-tunnel or predicted and flight data discussed in this report undoubtedly give the impression that we are extremely critical and/or in opposition to wind-tunnel testing. That is absolutely not the case. Wind tunnels have provided extremely good data for many years and are continuing to do so. As mentioned earlier, the wind-tunnel predictions of the X-15 aerodynamics were extremely good. There have been numerous aircraft developed that have had no aerodynamic discrepancies whatsoever. We at FRC resort to the wind-tunnel people continuously in support of our flight research, and they have bailed us out, so to speak, on many occasions. We always request wind-tunnel support whenever we make other than a minor configuration change. We are completely dependent on wind-tunnel predictions in many things that we do, such as air launch. We depend entirely on these predictions to assure that we have no collisions during separation.

We know, too, that the wind-tunnel people are not blind to their own limitations. We have supported, at their request, a number of combined wind-tunnel/flight-research correlation tests designed to improve their

[43] This is an apparent reference to Yeager's flight in the X-1A on 12 December 1953. Bell engineers had warned him before the flight that the aircraft might go out of control at speeds above Mach 2.3, but Yeager flew the X-1A to Mach 2.44 (1,612 miles per hour) despite the warnings, which proved correct. He shut off the rocket engine, but the aircraft became violently unstable, going into something like an oscillatory spin with frequent roll reversals. He was thrown about the cockpit as the X-1A lost altitude, falling some 50,000 feet (from an altitude of about 76,000 feet). Semiconscious, Yeager brought the decelerating aircraft into a normal spin, recovering to level flight at about 25,000 feet. Subsequently, he landed on Rogers Dry Lakebed. A pilot with lesser skills and instincts would probably have perished. See Hallion, *On the Frontier*, pp. 292, 308, and Hallion, *Supersonic Flight: The Story of the Bell X-1 and Douglas D-558* (New York: MacMillan, 1972), p.174.

predictive capability in areas where they know they have problems simulating the total flight environment.

One might ask, if we are not critical of the wind-tunnel results or people, what are we critical of? The answer is that we are critical of the system. The system is not closing the loop. When discrepancies are noted between wind tunnel and flight, they are seldom examined in sufficient detail to pin down the actual reason for the discrepancies. The tendency is to dismiss the discrepancy with an excuse that the tares were wrong, or that the model was not representative, or that propulsion effects were not duplicated, and so on. There is little enthusiasm to go back and prove it. Both wind-tunnel and flight people are more enthusiastic about moving on to the next program [than about investigating the problem on an existing or completed program]. As a result, we see reoccurring problems continuously.[44] Our track record hasn't improved that much except in such catastrophic problem areas as roll coupling and the T-tail deep stall problems. Significant aerodynamic discrepancies still show up as do performance- and control-related deficiencies.

One of NASA's primary responsibilities is to close the loop from the wind-tunnel to flight, and yet in many instances this has never been adequately done. An honest attempt to do this was made with the XB-70 in trying to explain the large discrepancy between predicted and actual performance. New models were constructed and a new series of wind-tunnel tests was conducted. These tests showed somewhat better agreement with flight results and yet a 10 percent discrepancy still existed; this still would result in a significant range discrepancy. No further attempts were made to improve the correlation. A more serious effort to correlate wind-tunnel and flight data is currently underway with the F-111 Transonic Aircraft Technology program.[45] Hopefully this will be carried through to its ultimate conclusion.[46] This, of course, does not provide the correlation we need at higher Mach numbers. We need additional validation of wind-tunnel predictions in the hypersonic speed region since the only good flight data is from the X-15. We have successfully flown some small-scale vehicles at hypersonic speeds, but the flight data obtained was minimal and compromised by lack of accurate air data. As far as environmental problems are concerned, we don't anticipate any significant new problems in the near future since there are no current plans for higher performance aircraft.

A potentially serious problem for future aircraft designers is emerging as the trend toward contracted wind-tunnel operation

---

[44] At this point in the text, Milt intended to insert a table listing some of the recurring problems, the aircraft involved, and the period of development of the particular aircraft. Unfortunately, he apparently did not leave behind such a table, and I am not competent to construct it.

[45] Following a great deal of wind-tunnel testing at NASA's Langley Research Center and by General Dynamics, the Flight Research Center began flight research with an F-111 on 1 November 1973. The program continued until the late 1970s and was resumed in a second phase in the mid-1980s. See Richard P. Hallion, *On the Frontier: Flight Research at Dryden, 1946-1981* (Washington, DC: NASA SP-4303, 1984), pp. 207-209.

[46] In a later document in the files he left behind, Milt wrote, "All objectives of the TACT program have been met in FY 1978. In general, flight data has validated the improvements to aerodynamic characteristics as predicted by wind-tunnel and calculated data." Draft for "Annual Report of Research and Technology Accomplishments and Applications, FY 1978, Hugh L. Dryden Flight Research Center," p. 3, held in the Dryden Historical Reference Collection.

increases. We certainly aren't going to produce experts like [Richard] Whitcomb, [Robert T.] Jones, [John] Becker, [Eugene] Love, [Alfred J.] Eggers, [Clarence] Syvertson, [probably Robert W.] Rainey[,] and on and on[,] with contracted wind-tunnel operations. The aerospace contractor will be pretty much on his own in assessing the quality of his data. This is not to say that an individual contractor cannot do an excellent job on his own. Individual contractors cannot, however, have access to all the data that NASA [does] because of proprietary problems. They thus would be hampered in developing equivalent experts. NASA has in the past provided continuity and good advice to many contractors in the development of new configurations. NASA has also had the luxury of looking at far-future and high-risk concepts. This is a luxury industry could not afford. Thus, this trend toward contractor operation of all wind tunnels should be halted and hopefully reversed.

To make the future look even more bleak, there is pressure from some sources to eliminate all government aeronautical research and development. In our opinion, that could mean complete disaster for the U.S. aerospace industry in the international sale of aircraft and would, as a result, significantly affect the country's balance of payments. We feel the present cost of government aeronautical research and development is a very minimal and essential subsidy to our aerospace industry. We in the flight-research business do feel that we will still be in business for some time to come because of the many other potential new problem areas alluded to earlier.

The Future

Flight research in the next five to ten years doesn't look as though it will be very exciting. As of now, there are no big advances planned in terms of flight-envelope expansion for future aircraft. There are no serious efforts to design a triple-sonic fighter. Even more disturbing is that there is no real enthusiasm for a hypersonic research aircraft. True, we have flown the X-15 to hypersonic speeds, but the X-15 did not address many of the critical disciplines such as structures, propulsion, etc. Kelly Johnson was pretty much on his own in designing the superb YF-12 and SR-71 aircraft. We are currently flying two of these aircraft in an attempt to determine why they fly as well as they do. So far we have seen several discrepancies between theoretical and actual data. The boundary layer conditions, for example, are significantly different from what one would predict.

Considering the number of aerodynamic, propulsion, and performance discrepancies we have observed, it is obvious that a lot of good, sound engineering judgment was applied in the design of the aircraft. It is again only after the fact that we are capable of explaining why. It is discouraging to realize that we have to resort to operational aircraft to obtain data to update theoretical and wind-tunnel predictive capability.

We feel we critically need a new series of research aircraft to stimulate new aircraft development. The early series of research aircraft stimulated a wide variety of new supersonic operational aircraft in the 1950s. The swept-wing F-100 was based on the success of the D-558-2. The straight-wing F-104 was based on the X-1 and X-3 successes. The F-102 was based on the X-4 and XF-92 results.[47] And finally, the F-111 stemmed from the marginally successful X-5 results. Subsequent to that extremely stimulating period, there have been no real[ly] imaginative developments in aircraft configurations.

---

[47] Milt had said XF-91. As an anonymous reviewer commented, this should be the XF-92. "The XF-92 was a delta wing. The XF-91 was a reverse taper prototype."

The current prototype fighter program is a step in the right direction, as are the Highly Maneuverable Aircraft Technology (HiMAT) and Advanced Fighter Technology Integration (AFTI) programs. However, none of these are real challenging. They are all still focused primarily on the transonic flight regime, as were the F-14 and F-15. To us this is indicative of short-sightedness. We feel that the Vietnamese conflict convinced too many advanced planners that all future combat would take place at transonic speeds. They have not acknowledged the fact that this was inevitable at the time because of the thrust-to-weight ratios of the aircraft involved. The new prototype fighters are capable of supersonic combat because of their high thrust-to-weight ratios and can virtually eat up the highly touted and transonically optimized F-14 and F-15, with properly developed tactics.

There are those doomsayers who say that we will never have supersonic fighter combat. That is ridiculous. Give a fighter pilot an aircraft capable of supersonic combat and he'll find a way to use that advantage, just as he did with the early jet aircraft against the best propeller-driven aircraft. You quickly learn not to fight in the opponent's best arena. If one believes that philosophy, then the Spad is still the best fighter ever conceived.[48] In the case of the F-14 and F-15, the philosophy is that if the Foxbat[49] comes down to my piece of the sky, I'll eat him up. If he doesn't, I'll shoot him down from below. If that philosophy holds true, the Navy should resurrect the old B-52s, hang a hundred or so Phoenix missiles on them and have a fleet of flying battleships. Better yet, rebuild some dirigibles and have worldwide air superiority.

If the U.S. is to retain air superiority, we must begin developing aircraft that can go up and stick their noses up the Foxbat's tailpipe. We have the technology in hand to build a true triple-sonic fighter that can maneuver aggressively in the Foxbat's arena. We can give our pilots the perch for the first time since World War I. We lost the perch in World War II and haven't regained it since. It's time we did.

---

[48] There were several models of Spads, but undoubtedly Milt is referring to the Spad XIII built in France at the end of World War I and flown by the French, Italian, and Belgian forces as well as the American Expeditionary Force. Many famous pilots flew it, including Captain "Eddie" Rickenbacker. The fighter served well into the 1920s in seven countries.

[49] The Foxbat was the NATO reporting name of the Soviet MiG-25, which could climb to over 123,000 feet.

About the Author

Milton O. Thompson was born in Crookston, Minn., on 4 May 1926. He began flying with the U.S. Navy as a pilot trainee at the age of 19 and served with the Navy for six years, including duty in China and Japan at the end of World War II. At the conclusion of his period of service with the Navy, he entered the University of Washington and graduated with a bachelor of science degree in engineering in 1953.

After two years of employment with the Boeing Aircraft Co. as a flight test engineer, Milt became an engineer at the High-Speed Flight Station, predecessor of NASA's Dryden Flight Research Center. He joined the Station in 1956 and was assigned to the pilots' office two years later, serving as a research pilot until 1968. Associated with many significant aerospace projects in that period, Milt is perhaps best known as one of the initiators of and pilots in the lifting-body program and as an X-15 pilot. He was the first pilot to fly a lifting body—the lightweight M2-F1—and later piloted the heavyweight M2-F2. He flew the X-15 14 times, reaching a maximum speed of 3,723 miles per hour and a peak altitude of 214,100 feet.

In 1968 Milt moved on to become the director of Research Projects. He followed that assignment with appointment in 1975 as Chief Engineer. He remained in that position until 1993, the year of his death. During the 1970s, he was a member of NASA's Space Transportation System Technology Steering Committee, in which capacity he was successful in applying the knowledge from the lifting-body and X-15 programs to the Shuttle design. That is, he led the effort to design the orbiters for power-off landings rather than increase their weight with air-breathing engines for airliner-type landings. His work on this committee earned him NASA's highest award, the Distinguished Service Medal—but one of numerous awards he received.

In the year before he died, Milt published *At the Edge of Space: The X-15 Flight Program* (Washington, DC, and London: Smithsonian Institution Press, 1992). After his death, Curtis Peebles edited and completed his *Flight without Wings: NASA Lifting Bodies and the Birth of the Space Shuttle* (Washington, DC: Smithsonian Institution Press, 1999), of which Curtis is listed as co-author. In addition, Milt wrote or co-authored about 20 technical reports.

Monographs in Aerospace History

Launius, Roger D., and Gillette, Aaron K. Compilers. *The Space Shuttle: An Annotated Bibliography.* (Monographs in Aerospace History, No. 1, 1992).

Launius, Roger D., and Hunley, J.D. Compilers. *An Annotated Bibliography of the Apollo Program.* (Monographs in Aerospace History, No. 2, 1994).

Launius, Roger D. *Apollo: A Retrospective Analysis.* (Monographs in Aerospace History, No. 3, 1994).

Hansen, James R. *Enchanted Rendezvous: John C. Houbolt and the Genesis of the Lunar-Orbit Rendezvous Concept.* (Monographs in Aerospace History, No. 4, 1995).

Gorn, Michael H. *Hugh L. Dryden's Career in Aviation and Space.* (Monographs in Aerospace History, No. 5, 1996).

Powers, Sheryll Goecke. *Women in Flight Research at the Dryden Flight Research Center, 1946-1995* (Monographs in Aerospace History, No. 6, 1997).

Portree, David S.F. and Trevino, Robert C. Compilers. *Walking to Olympus: A Chronology of Extravehicular Activity (EVA).* (Monographs in Aerospace History, No. 7, 1997).

Logsdon, John M. Moderator. *The Legislative Origins of the National Aeronautics and Space Act of 1958: Proceedings of an Oral History Workshop* (Monographs in Aerospace History, No. 8, 1998).

Rumerman, Judy A. Compiler. *U.S. Human Spaceflight: A Record of Achievement, 1961-1998* (Monographs in Aerospace History, No. 9, 1998).

Portree, David S.F. *NASA's Origins and the Dawn of the Space Age* (Monographs in Aerospace History, No. 10, 1998).

Logsdon, John M. *Together in Orbit: The Origins of International Cooperation in the Space Station Program* (Monographs in Aerospace History, No. 11, 1998).

Phillips, W. Hewitt. *Journey in Aeronautical Research: A Career at NASA Langley Research Center* (Monographs in Aerospace History, No. 12, 1998).

Braslow, Albert L. *A History of Suction-Type Laminar-Flow Control with Emphasis on Flight Research* (Monographs in Aerospace History, No. 13, 1999).

Logsdon, John M. Moderator. *Managing the Moon Program: Lessons Learned from Project Apollo* (Monographs in Aerospace History, No. 14, 1999).

Perminov, V.G. *The Difficult Road to Mars: A Brief History of Mars Exploration in the Soviet Union* (Monographs in Aerospace History, No. 15, 1999).

Tucker, Tom. *Touchdown: The Development of Propulsion Controlled Aircraft at NASA Dryden* (Monographs in Aerospace History, No. 16, 1999).

Maisel, Martin D.; Demo J. Giulianetti; and Daniel C. Dugan. *The History of the XV-15 Tilt Rotor Research Aircraft: From Concept to Flight.* (Monographs in Aerospace History #17, NASA SP-2000-4517, 2000).

Jenkins, Dennis R. *Hypersonics Before the Shuttle: A History of the X-15 Research Airplane.* (Monographs in Aerospace History #18, NASA SP-2000-4518, 2000).

Chambers, Joseph R. *Partners in Freedom: Contributions of the Langley Research Center to U.S. Military Aircraft in the 1990s.* (Monographs in Aerospace History #19, NASA SP-2000-4519).

Waltman, Gene L. *Black Magic and Gremlins: Analog Flight Simulations at NASA's Flight Research Center.* (Monographs in Aerospace History #20, NASA SP-2000-4520).

Portree, David S.F. *Humans to Mars: Fifty Years of Mission Planning, 1950-2000.* (Monographs in Aerospace History #21, NASA SP-2002-4521).

Those monographs still in print are available free of charge from the NASA History Division, Code ZH, NASA Headquarters, Washington, DC 20546. Please enclosed a self-addressed 9x12" envelope stamped for 15 ounces for these items.